Simple Patterns in Mathematics

— English —

True Patterns of Multiplication and Division Re-introduced to Students and Teachers

David T. Carrott, Ph.D.

Simple Patterns in Mathematics

2024

Multiplication & Division

This written work is a work of nonfiction. Unless otherwise noted, the author makes no explicit guarantees of the information contained in this written work. Care and cited references were used to ensure as much accuracy of the historical information.

ISBN: 978-1-958321-21-8 (English hardcover)
ISBN: 978-1-958321-20-1 (English softcover)
ISBN: 979-8-9876046-3-2 (English Kindle)

United States of America,
Library of Congress Control Number: 2023900921

Author Published revision 2024

Preface

My whole life has been an adventure of finding patterns in science and mathematics. Whether it was the simple patterns of the multiplication table, a pattern of electromagnetics, the geometric patterns in 2D imagery, or even patterns of surface texture as Fractals. It resulted in my finding answers where others followed the use of traditional teachings. I have been fortunate to be involved in leading-edge technology of computers (i.e., personal, optical, and quantum), navigation systems (i.e., inertial and laser), materials (i.e., piezoelectrics, glass-ceramics, concrete, and rubber), medical (i.e., ophthalmologic and ontology research), and machine-based learning (i.e., analog, optical, and digital-based target recognition or machine learning).

I have lectured on many of these leading-edge technologies by drafting my talks using the fundamental idea of knowing my audience. Whether it was to grade schools, high schools, academic or professional societies, senior military personnel, or Congressional representatives; I have used patterns to help them understand. I was once complimented by an Army Base Commander at an aerospace conference of corporations. In his closing remarks, he referred to me as the *evangelist of new technology*. I have never received such a verbal honor. In my later years, people frequently referred to me as a genius. My response has always been, "No, just persistent at what I do." If you don't know something, then pursue it till you do. Whether in school or at work, you are your best driving force.

Persistence can be forced on anyone, and what do you gain?
But, internal persistence can open many opportunities.

David T. Carrott

This book draws upon your first experiences in learning Arithmetic, consisting of the mathematical Arithmetic operators of addition, subtraction, multiplication, and division. Is it not strange that with all the complexity of the first computers, which are so powerful and fast; are based on the same foundation of all arithmetic operations. So, read about the patterns of the multiplication table described in Chapter 2, and you can be persistent in your endeavors.

Even if you already know things beyond simple Arithmetic, such as Algebra, Trigonometry, and Calculus; understanding the multiplication table's various patterns helps one understand many other patterns in mathematics and science. This book reintroduces an approach of "understanding" the multiplication table as patterns of numbers, which goes back 400, 800, and 1200 years; all all the way back to 3000 B.C. See Chapter 1.

Every 400 years, reference 0.1, someone adds a new piece to each prior 400-year discovery. My added part came in 2000 which included: Curriculum Roadmap of Math, Rote vs Pattern learning, Patterns of the multiplication table, and a modern-day adaptation of John Napier's Bones that now includes Division.

I have been persistent in finding patterns in science and mathematics that have improved my value in the trade. I believe with these patterns shared in this book; you too can enhance your understanding, value, and future to others.

This book is divided into the two lead sections of the six.

The **first section** sets the foundation through a historical review, methods of learning, and this new understanding of patterns within the multiplication table. This version includes corrections and improvements, by splitting Learning Methods and Patterns into Chapters 2 and 3.

Chapter 1: Introduction which sets the foundation for the other three major sections: Multiplication Table, Applied Multiplication, and Applied Division. This section helps the reader understand the question finally answered in Chapter 2. "Why do we need to memorize the multiplication tables of zero (0) through twelve (12)?"

Chapter 2: Learning Methods are introduced to help the reader understand several methods of learning: Rote, Meaningful, Critical, and Patterns. Each is presented with its strengths or advantages. However according to this author, only one prevails with all the best approaches: Patterns.

Many teachers support the idea of memorization, which is used in Rote. However, it is proposed that this particular learning method has been the downfall of our education system for math. This one approach has driven the United States education system of mathematics and science to its current fallen state.

The key to the success of the next Chapter, Patterns, is seen in asking any grade school student why they "Hate" math.

Chapter 3: Patterns are introduced to help you know the multiplication table without memorization. If you already memorized them that is fine, but did you know that there are patterns to this simple table? These patterns draw upon the simple concept that multiplication is a shortcut, or repeated pattern of addition, which is the same number added 0 or more times. With these patterns, you can better understand multiplication and other math principles, without memorization. Many of these patterns are taught later in higher levels of Mathematics, but they can help you to begin to understand them now.

These patterns are arranged in four groups that represent:

- Column-based Patterns (Patterns 1-5)
- Table-Based Patterns (Patterns 6-10)
- Digit-Based Patterns (Pattern 11)
- New Patterns (Pattern 12-17)

Organizing patterns like this opens opportunities to clarify what you are doing, but it also opens opportunities to define new or undiscovered groupings or patterns. As I am sure there are other patterns yet to be defined, I submit a challenge to identify other patterns not yet listed.

The **second section** covers the traditional concepts of Tally-x, reference 0.2, and multiplication and division. This section represents the major update in this book version, where Chapter 6 is changed to Decimal Math. The original Chapter 6 was Squares and Cubes, but that will be covered in Algebra.

Chapter 4: Multiplication is where the concept of single and multi-digit multiplication is introduced. It starts with a review of traditional "long multiplication" and then applies the multiplication patterns to the "long multiplication" approach. With this, you can start with single-digital multipliers (SDM) and then grow that into two, three, and larger multipliers. This is then compared to the Tally-x approach to multiplication, so you can discover that it is easy to work on simple SDM problems, as well as, a multiple-digit multiplier (MDM). This MDM helps you understand the "long multiplication" approach, making your teacher happy. And no calculators or batteries are required.

Chapter 5: Division is reintroduced here by reviewing the traditionally taught "long division" approach and then applying the multiplication table patterns to the original "long division" approach. You start with a single-digital divisor (SDD) and then grow that into two, three, and large dividers. You will find this approach is just as easy to do and understand then multiplication.

The proof of this approach is if you ask any student today, which of the four arithmetic operators they "hate" (that is, addition, subtraction, multiplication, and division), most of the time the answer is division. But, ask the students who know these simple multiplication patterns, and the answer is none of them because they understand it.

Chapter 6: Decimal Mathematics shows the use of Tally-x sticks for the multiplication and division of integers, which then shows the simplicity of doing decimal multiplication and division. This approach expands the idea of the positional value system, through a simple decimal pattern that separates the integer from the subinteger values. This is the next step in learning multiplication and division, understanding decimals, without the complexity normally taught.

The final **Index section** covers a selection of keywords and terms found throughout the book, that I thought would be helpful. However, it was not found to be as affected in non-English translations. As such, it is only available in the English version. The Index provides the reader with a quick definition and/or where a subject is covered in the text using chapter and page locations.

For example, the term Napier can be found in Section 1.2.9: 1600 AD and in many other places. I have a great admiration of John Napier of Edinburgh Scotland in the United Kingdom, because of his genius and approach to multiplication. There are many books about his life beyond that discussed in this book.

Figure 0.1: Merchiston Castle - home of John Napier.

I would also encourage you to visit Edinburgh Napier University - Merchiston Campus and John Napier's original family home. See Figure 0.1. Stop by the university information office and inquire about a tour. Just be careful, the doorways and stone spiral steps are smaller than today. It is a worthy and quite a unique opportunity.

I was fortunate to visit and tour the Merchiston campus and residence of John Napier in September 2022. It was one of my greatest adventures; as it brought this 16th-century math scholar to life for me. It is one thing to read about someone, it is altogether something else to visit their home and place of study. Merchiston Castle and the University campus is only 1.5 miles from downtown Edinburgh and Edinburgh Castle and well worth the walk.

What you see in Figure 0.1 is from the south side and the student area. This shows the upper three floors of the castle. But, there is a cellar below ground with a door in the back that comes out under the glass walkway of the University. Imagine going to a university that is built around a castle that dates back to 1454, *reference 0.3*.

The doorway on the left, in the figure, is raised above the student quarter area. There used to be a L-shaped stairway leading up to the door. This raised door made it easier to defend the castle against groups that troubled the countryside during these early years.

Forward

Math was not a strong point for me in my early school years. It wasn't until High School Geometry that I found something substantial that I could get my mind around. However, during three years of High School, I attended three different schools in two states and each had a different academic viewpoint. This confusion was a challenge that I was willing to explore and it took years to finish.

It was in college that I started to understand why I was not getting my mind around math. Trigonometry was hard, Calculus was almost impossible, and statistical math was a total loss — fortunately, my persistent nature in finding patterns in things was what helped me.

In the 1990s, I had the luxury of finally finding what I thought was the answer to understanding Math. This involved a series of events that started explaining how a slide rule worked to some friends. Pocket calculators replaced slide rules by the late 1970s. The slide rule became forgotten and ended up in the back drawers or the trash. I showed people how simple it was to make them out of popsicle sticks. The linear scales provided a mechanical tool for adding and subtracting numbers, and logarithmic scales were the power that allowed adding (and subtraction) for multiplication (and division), as well as using other scales to do even more functions of math.

It was the slide rule that introduced me to the father of logarithms, John Napier. Although his understanding of logarithms was impressive, it was his other tool that attracted my attention: Napier's Bones. I could finally show people how to do multiplication in a more natural and understandable method. However, it soon turned into much more, including this book, reference 0.4.

My demonstration of the slide rule and Napier's bones, started in the early 1990s at NASA Dryden Flight Research Center, now the Armstrong Flight Research Center, when I was asked to fill in for a sick NASA engineer for an Engineering Week presentation at a local Grade School. Word soon spread to other schools about my visit. Soon NASA was being asked by other schools to have me do the same presentation at their schools. This continued for three years, and at each school, I improved my presentation material, and I looked for more information. I became popular at these presentations and received many notes and letters from classes and teachers. I was reintroducing John Napier and a new way to do math to the local schools.

After moving my family to Virginia in 1999, I started teaching at the local Junior College in the evening. It was during this time that I developed the Carrott Map and how it influenced the curriculum of Mathematics. During this time, I was researching how John Napier (1600s) came up with the idea of Napier's Bones. It took another nine years to make the historical connection to the lattice method which was originally documented by Fibonacci (aka. Leonardo Pisano or Leonardo de Pisa, 1200s). The lattice method is commonly taught in local schools to help students see the place value method of multiplication. Then two years later, in 2010, I made the connection to Hindu, Vedic, and earlier math materials going back to 3,000 B.C.

It seems Mathematics is all about shortcuts in math that were discovered and continue to be re-discovered by empirical observations of patterns in nature. There is a simple un-written math idea: where using a shortcut in athletic competition is a foul, discovering shortcuts in math is rewarded. May this book help you find out some of the patterns and alternatives that define math?

I would at this time like to thank my wife Sandy for her support over the years in my life's work and also for the inspiration to "keep moving forward." It is a pleasure to share these patterns of multiplication with the world and also help others to "keep moving forward."

Table of Contents

Prefix and Forward Endnotes.

0.1 This is not the conventional approach taught in schools today; however, the approach has been taught in the 400-year cycle for over 5,000 years. In some respects, it is the approach still taught to students outside the United States when used with an Abacus.

0.2 Disney, Walt, “Around here, however, we don’t look backward for very long. We keep moving forward, opening up new doors and doing new things, because we’re curious and curiosity keeps leading us down new paths.”

0.3 “Merchiston Tower” also known as Merchiston Castle, Wikipedia. This site shows the L-shaped staircase coming out into the University student area in 2012.

0.4 John Napier made significant contributions that changed mathematics during the 16th and early 17th centuries. These include: 1) invention of Logarithms, 2) Napier’s Bones, 3) Decimal Notation, and 4) Trigonometric circular functions.

Chapter 1: Introduction

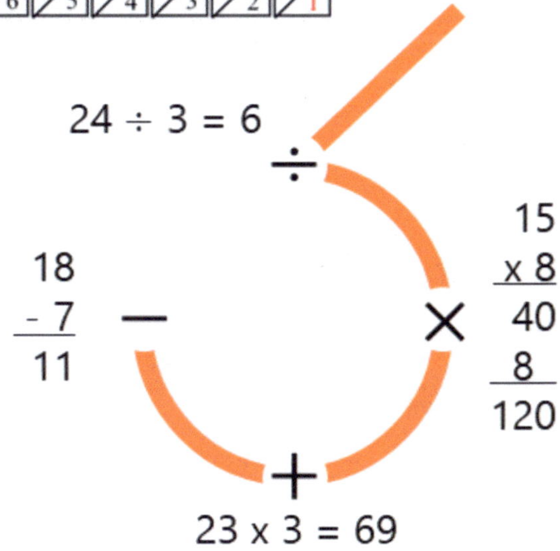

1.1 Introduction.

Mathematics is all about shortcuts that are based on empirical observations of patterns in nature. There is a pattern in all things of nature and God. The Lord said, "My house is a house of order … ", *reference 1.1*, which implies it is not a house of confusion. If we pattern our homes and lives according to God's guidance, we will be better in our homes and our lives.

1.1.1 Pattern of Numbers.

The natural world contains an infinite variety of patterns. Patterns are found in plants, foliage, and animals. All living things generate patterns. Patterns are continually being discovered using simple physical laws. There are patterns in the sand created by blowing winds. There is a pattern in the vortex of a whirlpool and the formation of an ice crystal.

In mathematics, there is a science of Chaos called Fractals. Fractal science is a fascinating mathematical understanding of graphic imagery that gives us a new way to think about the world. Benoit B. Mandelbrot in 1975 coined the term "fractal" from the Latin word *fractus*, meaning irregular or fragmented. He spent his life shocking the science establishment and is now revered for inventing a new geometry. Fractals were used in the 1980s to measure surface roughness, including runway surfaces for NASA Landing System Research Aircraft (LSRS, study of Shuttle tires in the 1990s), *reference 1.2*.

However, even simple mathematics such as Arithmetic operators (addition, subtraction, multiplication, and division) with their affected numbers have patterns. Moreover, that is what this chapter plans to show you. Figure 1.1.1-1 shows you the pattern of place values, which is one of the crucial features of a useful modern math tool.

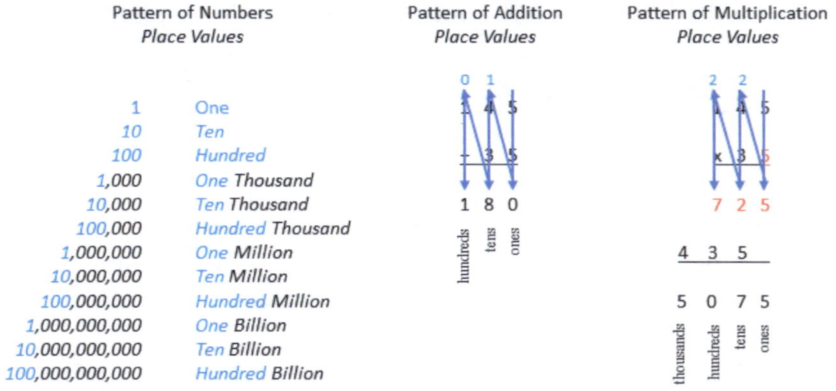

Figure 1.1.1-1: Place Values.

Place values are an essential tool in an effective math system that supports the efficient use of numeric symbols. Unfortunately, this was not the case in early European or Western cultures that relied on the Roman numeral system of values and symbols. However, it was supported by the Hindu-Arabic system of the place value and numbers 0 to 9. See section 1.3.

1.1.2 10-digit Numbers and Order.

The modern 10-digit numbering system is possibly the most simple and basic number system developed by humankind. It naturally comes from our 10-fingers. The problem is that we are taught to count by the same ten-finger approach, counting from 1 to 10. Moreover, it is considered part of the first sequence of numbers. See Figure 1.1.2-1 below.

1	2	3	4	5	6	7	8	9	10
11	12	13	14	15	16	17	18	19	20
21	22	23	24	25	26	27	28	29	30
31	32	33	34	35	36	37	38	39	40
41	42	43	44	45	46	47	48	49	50
51	52	53	54	55	56	57	58	59	60
61	62	63	64	65	66	67	68	69	70
71	72	73	74	75	76	77	78	79	80
81	82	83	84	85	86	87	88	89	90
91	92	93	94	95	96	97	98	99	100

Figure 1.1.2-1: Traditional Listing of Numbers.

The traditional arrangement of numbers, depicted in Figure 1.1.2-1, begins with the units column starting from 1 on the left, followed by 2 in the second column, 3 in the third column, and so forth. The rightmost column represents the tens place, with a unit value of zero (0) and a tens value of one (1).

However, this arrangement complicates the sequence of numbers compared to the original teaching method, often confusing rather than clarifying for students. Let me explain further. In the first row, the sequence appears as:

1, 2, 3, 4, 5, 6, 7, 8, 9, 10

However, if we include the tens placeholder in each number, denoted by the last value, it reads:

01, 02, 03, 04, 05, 06, 07, 08, 09, 10

In mathematics, we typically disregard any leading zeros in sequences like the one above, to emphasize the significance of place value. From this adjusted sequence, it's evident that the final value, 10, doesn't belong to the row of single digits. If we revise the figure by moving the value 10 and all subsequent values of that last row to the first column and shift them down by one row, as shown in Figure 1.1.2-2, and introduce a zero (0) value in the first row and column (i.e., zero), then start the ones place value sequence.

1s values

10s values

0	1	2	3	4	5	6	7	8	9
10	11	12	13	14	15	16	17	18	19
20	21	22	23	24	25	26	27	28	29
30	31	32	33	34	35	36	37	38	39
40	41	42	43	44	45	46	47	48	49
50	51	52	53	54	55	56	57	58	59
60	61	62	63	64	65	66	67	68	69
70	71	72	73	74	75	76	77	78	79
80	81	82	83	84	85	86	87	88	89
90	91	92	93	94	95	96	97	98	99

Figure 1.1.2-2: Proper Listing of Numbers.

Figure 1.1.2-2 shows how a number sequence should be, starting with zero in the first column. This helps the Place Value system form our 10-digit numbering system. Not only does it explain the Place Value, but it also eliminates one other issue from Figure 1.1.2-1. The last value in Figure 1.1.2-1 was 100, which does not fit either the row sequence or does it fit the last column sequence from top to bottom as Figure 1.1.2-2 does.

There are three reasons the proper listing of numbers is better than traditionally taught in the first year of schooling: Clairty in the Place Value, Logical and Coherent Structure, and Solid Mathematical Foundation.

This is seen in the following three applications:

1) … matches up to the place value system of 0-9, or 1s and 10s. The advantage of this is that it allows for all future place values, i.e., 100s, 1000s, etc.
2) … shows the diagonal matching of 11, 22, 33, 44, …, and 99. Whereas, Figure 1.1.2-1 does not, i.e., 1, 12, 23, 34, …, 89, 90 to 100.
3) … supports future mathematical understanding and patterns found in Chapter 3.

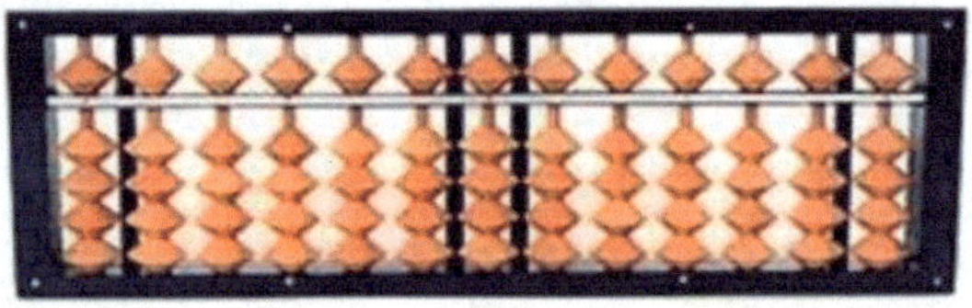

Figure 1.1.2-3: Soroban.

The Japanese Soroban, as shown in Figure 1.1.2-3, *reference 1.3*, directly extends from the proper listing of numbers where each column, similar to the rows in Figure 1.1.2-2, can only count up to 9. Understanding this type of place value system enables students to grasp the various practical applications of the abacus (*reference 1.4*) worldwide. Refer to sections 1.2 and 1.4 in this chapter.

The Soroban restricts each rod to counting up to nine, which reinforces the 10-digit numbering system depicted in Figure 1.1.2-2. Even the Romans adopted a similar understanding of place value, using symbols like "X" (ten) and "C" (100). Like the Soroban in Japan, the Romans aimed to simplify counting with a base-5 system using symbols such as "V" (5 counts), "L" (50 counts), and "D" (500 counts). Refer to section 1.3.

1.2: History of Multiplication.

The search for understanding the fundamental operations of mathematics began with the quest for improved methods of multiplication. John Napier and I both shared a passion for discovering more efficient multiplication techniques. My journey began in the late 1990s when I uncovered and comprehended John Napier's 16th-century bones. This concept was utilized to introduce an alternative multiplication approach to schools in the Antelope Valley of California during the 1990s, spanning over three years with approval from NASA's Armstrong Flight Research Center, formerly known as Dryden Flight Research Center (FRC), Public Relations office, and through word of mouth among local public schools, *reference 1.5*.

The documentation of John Napier's mathematical 'bones' remains a historical curiosity. There is some resemblance to the lattice multiplication method practiced in Arab and Turkish schools, which serves as an alternative to traditional long multiplication. Many teachers are acquainted with and teach the lattice method, often associating it with the publication of Leonardo Pisano's book "Liber Abaci" (Fibonacci in 1202 AD).

Wikipedia's article on John Napier states, "Napier made further contributions [in mathematics]. … Arab lattice multiplication, used by Fibonacci, was made more convenient by the introduction of Napier's bones, a multiplication tool using a set of numbered rods." The advantage of Napier's approach over that of the Lattice method was that the Lattice required a pre-knowledge of the multiplication tables of 0 to 9. Whereas Napier's approach had the calculations pre-built into the bones or numbered rods. Both methods formed a type of calculator, but by adding Napier's pattern approach we remove the needed memorization or pre-knowledge and its use is unlimited as a user's calculator. For more details see Chapters 1.2.2 - 1.2.5 and 1.2.8 - 1.2.9.

The following is a summary understanding of the history of mathematics associated with 0-9 numbers, multiplication, and division in a corrected chronological order. This history goes back to the source of our 0-9 numbering system and included, to many people's surprise the basic principles of arithmetic, algebra, geometry, trigonometry, and calculus. From this one place math and its concept went east to Chinese and then Japanese. It also went west to the Middle East and on to Europe, including India (Hindu), Babylon, Egypt, Greece, North Africa, and then to Europe. But, our present-day understanding is limited to European and Greek math history. Let's start with the earliest known math history in 3,000 A.D.

1.2.1 3,000 BC.

Mohenjo-Daro was one of the largest cities of the ancient Indus Valley Civilization (IVC), see Figure 1.2.1-1, *reference 1.6*. The civilization spanned much of what is now Pakistan and Northwestern India, with settlements spread as far west as the Iranian border in Bactria, and as far south as the Arabian Sea of western India in Gujarat. The archeological discoveries at Mohenjo-daro revealed that as early as 3,000 BC, the inhabitants of the land of Sindhu – the Hindus – built and lived a highly organized and advanced life.

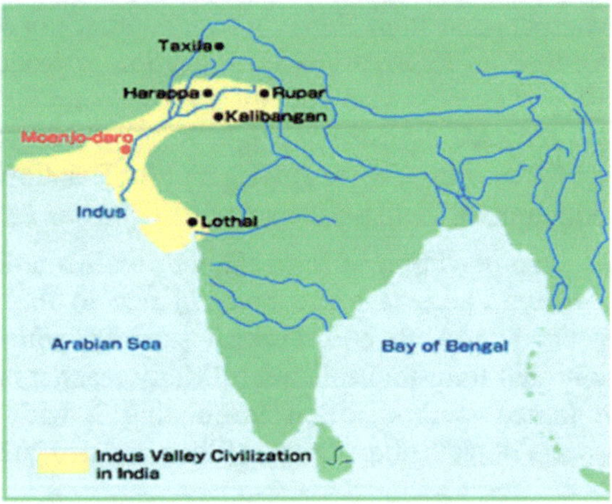

Figure 1.2.1-1: Indus Cultural Valley map.

This is the earliest rendition of the Hindu-Arabic system of numbers, less the Arabic symbol for zero. Here we find place values and 10-digit numbers. These early methods of a more natural approach to mathematics have been passed down and through the history of many cultures; including the Vedic math system, and its influence on Babylonia, Persia, Egypt, Turkey, China, and Greece.

1.2.2 1000 BC.

Valmiki, the father of Sanskrit, wrote the Ramayana, one of the two major Sanskrit epics, that forms the Hindu history or "Itihasa." The corpus of Sanskrit literature encompasses rich traditional math of poetry and drama as well as scientific, technical, philosophical, and Hindu religious texts. Vedic mathematics is a listing of sixteen sutras, or word formulae, in Sanskrit. Sutras describe the way the mind works and are therefore a great help in directing an appropriate method of solution. For example, the Weaving sutra is the heart of "Lattice" multiplication. From this, one can see that math is all about everyday patterns of life.

1.2.3 Kapâta-Sandhi.

The method of lattice multiplication traditionally used today is described in the Ganita-manjari, references 1.7 and 1.8, from the 16th century. Early

Indian mathematicians of the Indus Valley Civilization (IVC) in present-day Pakistan used various intuitive techniques for multiplication, including small slate tablets covered in dust, an early form of the abacus known by various names.

In Figure 1.2.3-1, the Lattice multiplication method shows a simple geometric demonstration of 256 times 45 in table form. It is drawn up with the columns (top side) labeled by the multiplicand, and the rows (left side) by the multiplier. Each box of the table is divided diagonally as a triangular lattice, top left to bottom right. The entries of the table hold the partial products, like decimal numbers. The outcome could be formed by summing down and right along the diagonals of the lattice. Note that the diagonal direction is different from the Gelosia diagram (section 1.2.4) and that the 45 is written upwards. Thereby the zero value of both values (multiplicand and multiplier) converges at the top left corner.

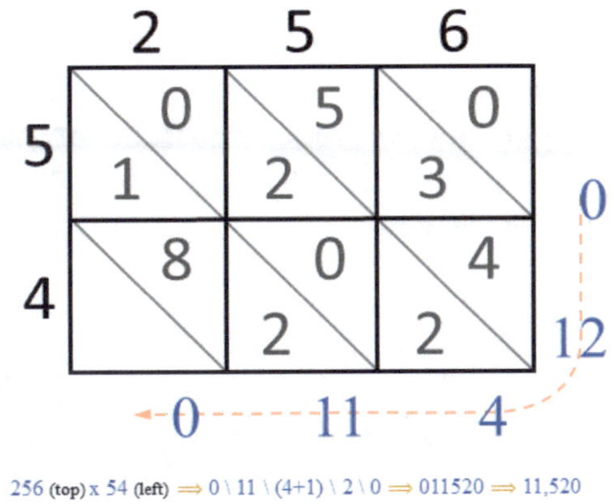

Figure 1.2.3-1, The Kapâta-Sandhi Method of Multiplication.

1.2.4 Gelosia.

The more traditional Lattice multiplication method used today is described in the Ganita-manjari, *reference 1.9 and 1.10*, (16th century), the early Indian mathematician of the Indus Valley Civilization (IVC), in present-day Pakistan, used a variety of intuitive tricks to perform multiplications using small slate tablets covered in dust, an

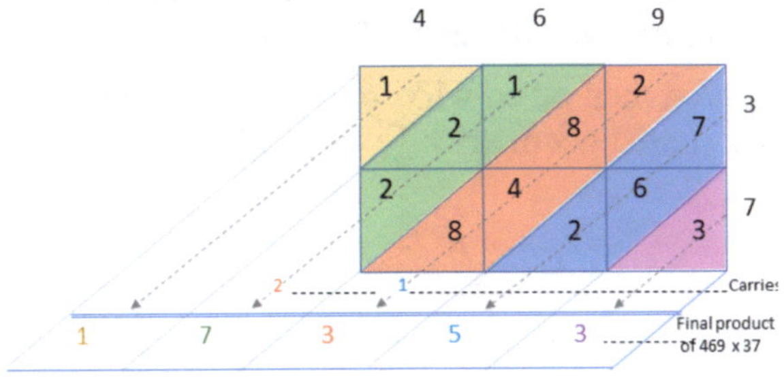

Figure 1.2.4-1: Lattice or Gelosia Method of Multiplication.

early version of the Abacus, and also called to by other names.

In Figure 1.2.4-1, the approach to multiplication is the earliest version of the Gelosia (see Leonardo de Pisa, section 1.2.8). It is also an early understanding that geometric shapes were used to prove the basic arithmetic operator.

In Turkey, at the Ottoman Palace, founded in 1455 AD, was the Enderun School, *reference 1.11,* where the gifted were taught math and science of the time. This included teaching the Hindu-Arabic numbers and the Matraki multiplication method, Figure 1.2.4-2, as part of their science and mathematics curriculum.

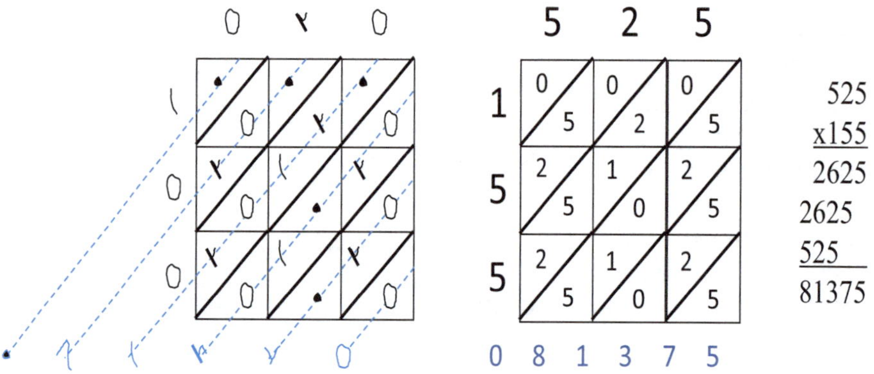

Figure 1.2.4-2 Matraki's method and Lattice method. for 155 times 525 = 81375.

You will note that in Figure 1.2.4-2, the solution follows the diagonal lines down and to the left, but the multiplier is still on the left side, as in the kapâta-sandhi and lattice methods. However, the Matriaki method shown in most historical documents is incorrect. The multiplier on the left side is displayed incorrectly. It is displayed as 001 (551) instead of 100 (155). A check of the inner column-row solutions will verify this error, *reference 1.12*.

Why was this lattice multiplication method so widespread and accepted? Because it could also be demonstrated geometrically, as can be shown in Section 1.2.5.

1.2.5 Geometry Math.

Both the kapâta-sandhi and Gelosia are also similar to a geometric area summary; however, they are in numeric form. Geometry as taught by the Greeks is the physical proof of multiplication. As addition is the linear sum of two values, multiplication is the geometric sum of two numbers.

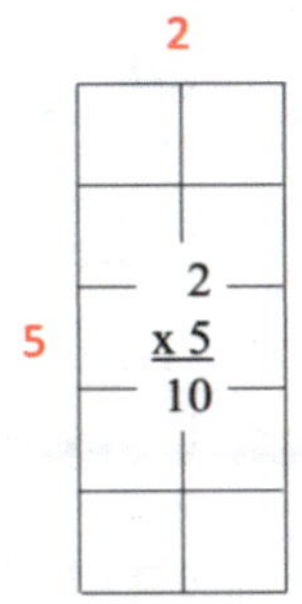

Figure 1.2.5-1: 2 x5 Pavers using Geometric Multiplication.

Using the same kapâta-sandhi values in the Lattice figure above, 256 times 45 can be expressed in a simple set of "square" pavers of a unit dimension. In Figure 1.2.5-1, the example, 5 times 2 would look like two sets of columns of five rows or 10 squares. So, the answer of 5 times 2 is 10.

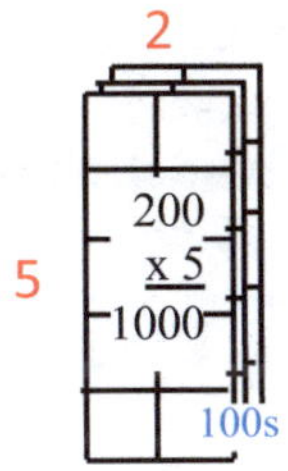

Figure 1.2.5-2: 200 x5 Pavers using Geometric Multiplication.

In Figure 1.2.5-2, we change the 2 for 200, then two blank placeholders (zero value) would line up behind the Figure. So, each blank square represents the place value or shift of the value to the left by the number of zeros. 200 has two zeros, i.e., 2 times 100, so the answer is 5 times 2 times 100 = 1,000.

Getting back to the original problem of 256 times 45 would be geometrically expressed and shown in Figure 1.2.5-3.

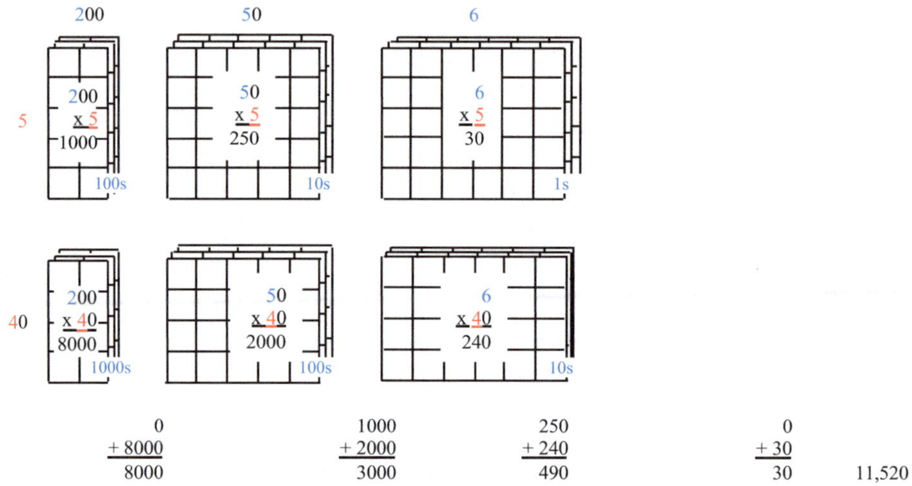

Figure 1.2.5-3: Geometric Multiplication Representation.

The Geometric method is similar to the kapâta-sandhi method, however, instead of a numerical approach, you are using square pavers. Much like using an abacus to do addition. As seen above the horizontal blue stroke arrow shows each paver grouping diagonally, with the sums shown along with the diagonal blue stroke arrows. Then all the diagonal values, with their correct place values in place, sum to the value on the right. This results in the answer of 11,520 units.

1.2.6 400 AD.

The development of the positional decimal system originates from Indian mathematics during the Gupta Empire period (320 to 550 AD). The Bakhshâlî manuscript is an ancient Sanskrit mathematics textbook dating back to this period in India, written on birch bark. The manuscript numbers are shown in Figure 1.2.6-1. This manuscript is perhaps the oldest in Indian mathematics and the earliest known to use a zero symbol. The Bodleian Libraries, University of Oxford, *reference 1.12*, holds the only existing set of the Bakhshâlî manuscript.

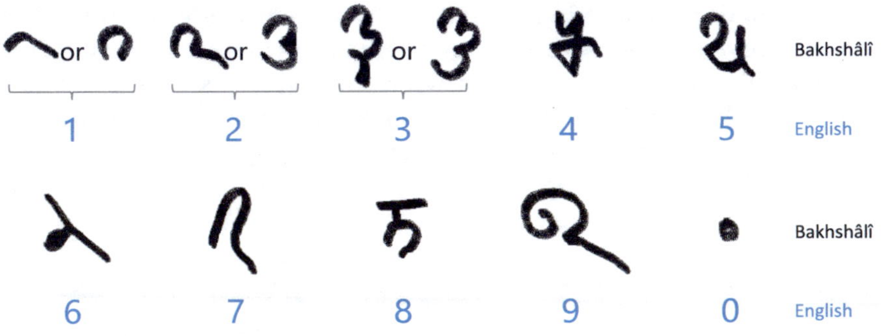

Figure 1.2.6-1: Bakhshâlî manuscript numbers.

The Bakhshâlî manuscript was found by a farmer in 1881 AD, buried outside a village called Baksheesh, in what is now a region of Pakistan. It remained an un-deciphered curiosity for many years until it was sent to the leading Indologist Augustus Frederic Rudolf Hoernlé. He deciphered it and showed it to be a portion of a lost ancient Indian arithmetical treatise, reference 1.13. He later retired from his Indian office in 1899 and continued his work at the University of Oxford, where his collections are on display at the Bodleian Libraries.

1.2.7 800 AD.

The development of Indian numerals, the place value system, zero, and the basic operators (addition, subtraction, multiplication, and division) were adopted in Islamic mathematics in the 8th century, as documented in al-Qifti's Chronology of the scholars (13th century). Figure 1.2.7-1 is one of several international stamps honoring Muhammad ibn Musa al-Khwarizmi (c. ~780-850 AD), a Persian mathematician and astronomer from the Khorasan province of present-day Uzbekistan. His arithmetic methods were crucial in disseminating the Hindu decimal place value system to the Islamic world and eventually to medieval European nations. See Section 1.2.8.

Figure 1.2.7-1: Muhammad al-Khwarizmi, 1983 Soviet Stamp.

The numeral system came to be known to both the Muslim/Arab mathematician Al-Khwarizmi, who wrote a book of Calculation using the

Hindu Numerals in about 825 AD, and the Arab mathematician. You can find books and movies about his life; however, I would venture that many are made more colorful than reality. Al-Kindi wrote four volumes on the use of the Indian numerals around 830 AD. These books are principally responsible for the diffusion of the Indian system of numeration throughout the Islamic world and ultimately also to Europe. The first dated and undisputed inscription showing the use of zero as a symbol is at Gwalior, 876 AD.

1.2.8 1200 AD.

Figure 1.2.8-1: Leonardo Pisano.

Leonardo de Pisa or Leonardo Pisano, also known as (aka) Fibonacci…", *reference 1.14*, was born in Pisa in Italy around 1170 AD, see Figure 1.2.8-1. His father held a customs position that involved his traveling to Northern Africa for Pisa market exchanges. Leonardo traveled with his father frequently and consequently received a Moorish education along with his traditional European education. His Moorish education introduced him to the Hindu-Arabic numbers, compared to his home's Roman numerals. His alternate name Fibonacci, is short for *filius Bonacci* from Latin *filius* ("son") + Italian *Bonacci* or Latin *bonum*, meaning "son of Bonacci" or "son of the good-natured [family]."

Later, he traveled about the Mediterranean visiting Egypt, Syria, Greece, Sicily, and Provence. Meeting with scholars and becoming acquainted with the various arithmetic operators used by the merchants.

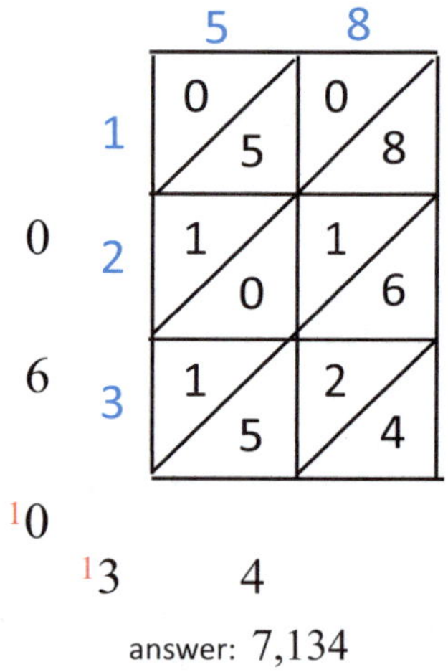

Figure 1.2.8-2: Lattice Multiplication Method.

Leonardo realized the math advantage of using Hindu-Arabic numbers over the traditional Roman numerals, which stimulated him to write his first of many books on mathematical methods, Liber Abaci. The book was later translated into English by Laurence Sigler, *reference 1.15*.

One of the methods of math he introduced in his first book was called the Gelosia method (see 1.2.4) or as we call it today the Lattice system. This method was learned from Moorish merchants. Gelosia is the Latin term for 'grid' as used on windows or Lattice. It gets its name from the fact that to do multiplication, you would fill in a grid that resembles a lattice one might find ivy growing on. This method has reappeared in schools and requires the memorization of the multiplication tables, 0 through 9. It is a method of multiplication and not a calculating device.

The Lattice method always starts with a blank Lattice line structure, which must be filled out by the multiplicand and multiplier. Then each diagonal square must be calculated by a Rote approach, see section 2.2, and the answer is written out. This method is no different from the long multiplication method. And like the multiplication method, both are limited to the multiplication function. The problem with the Lattice method is that the user must know the multiplication tables of 0 through 9, to fill in the initially empty lattice, which is then summed diagonally to resolve the final answer.

The Lattice multiplication method has a history that has gone unchanged, or slight changes, for over 3,000 years. In 1600 AD, that changed.

1.2.9 1600 AD.

John Napier was born around 1550 AD in Edinburgh Scotland. He was the 8th Laird (i.e., Lord) of Merchiston and was responsible for the lands and animal life under his care. He is considered a Scottish mathematician, physicist, astronomer, and astrologer, as well as a man of religion.

As a mathematician, he is best known for introducing the world to decimal notation, the creator of logarithms, and the father of Napier's bones. Just before he died in 1617 AD, he wrote the book Rabdologiae, *reference 1.14*, and described in great detail the workings of Napier's bones. It has been since translated from Latin to English as part of the Charles Babbage Reprint Series.

There are numerous books on the life of John Napier; however, his education exploits after attending the University of St Andrews are not as well known.

> "… John Napier did not enter schools until he was 13. He did not stay in school very long, however. It is believed that he dropped out of school in Scotland and perhaps traveled in mainland Europe to better continue his studies and little is known of those years …", *reference 1.16.*

Figure 1.2.9-1 Napier's Bones.

However, it can be drawn upon that his education in Europe, including the reading or understanding of the textbooks such as the Libra Abaci; might have provided John Napier with some insightful ideas. That if one looked at the Gelosia (Lattice) method, one could pre-load each 0 to 9 column of the lattice box, thereby creating a self-contained multiplication table for each digit column, as in Figure 1.2.9-1.

Correct version

Sides	A	B	C	D	
1	0	2	9	7	18
2	1	3	8	6	18
3	2	6	7	3	18
4	3	0	6	9	18
5	4	1	5	8	18
6	5	7	4	2	18
7	6	4	3	5	18
8	7	8	2	1	18
9	8	9	1	0	18
10	9	5	0	4	18
	45	45	45	45	

Figure 1.2.9-2 4-sided Bones.

Figure 1.2.9-2 shows a 4-sided selection of the multiplication sticks, and for each stick, the opposite value of the stick number is the 9-compliment, or value that sums the stick values to 9. I.e., the 0-stick has a 9-stick on the opposite side; the 5-stick has a 4-stick on the opposite side, and the 3-stick has a 6-stick on the opposite side. The opposite of each of these examples is also true for the 9-stick, 4-stick, and 6-stick. This clearly applies to any 2-sided stick version.

John Napier ingeniously applied it to his four-sided stick set, so that each stick value is depicted 4 times. In some products of John Napier's 4-sided stick, they got the values wrong, such that some numbers are shown more than others. The correct version has all numbers represented 4 times, as in Figure 1.2.9-2. Note that the rows add to 18 (double of 9 as it should) and the columns add to 45 (as it should, (0+9) +(1+8x) + (1+7) + …+ 5).

Moreover, the side-by-side placement of any two or more multiplication tables would generate a universal form of *Gelosia* (Section 1.08 to 1.13). This insightful idea was the genius of John Napier, where the memorization of the multiplication tables is not required. It is not a slight change in the lattice multiplication method; it is something new and it is genius. John Napier had discovered a calculating tool that was portable and easy to use. *He took a long-*

excepted mathematical multiplication method and turned it into the first general-purpose numeric-value decimal-based calculator. As a calculating tool, it was also universal enough to allow any two single or multi-digit numbers to be multiplied. But, it was also general enough that could do more than multiplication; it could do division, square roots, and cube roots. This was written by Napier as "Rabdology" before he died in 1617 AD but wasn't published until 1618 AD, *reference 1.17*. Napier coined the word rabdology from Latin for *rod* and *calculation* to describe his method of performing arithmetical operations employing rods or (ivory) bones.

Napier's bones are the foundation of Tally-x with its advantages over the Lattice (or Gelosa) system in that memorization of the multiplication tables is not necessary. John Napier expanded the simple geometric genius of the Lattice multiplication system into a full set of vertical sticks for each value, 0 through 9. Thereby reducing any risk of mistakes and opening up a new gateway for mathematics. You could say he invented the first "Pocket Calculator" or traveling calculator.

This is why memorization is not necessary; it is a pattern of the multiplication of numbers. For example, with its frequent use, it becomes intuitive as it has too many students and parents alike. Math is not that complicated, and we can demonstrate a simple example using the same Kapâta-Sandhi multiplication example from Figure 1.2.9 -3.

Step 1: Remove the necessary multiplicand sticks and appropriately set them up across the top as the *multiplicand* Figure 1.2.9-3.

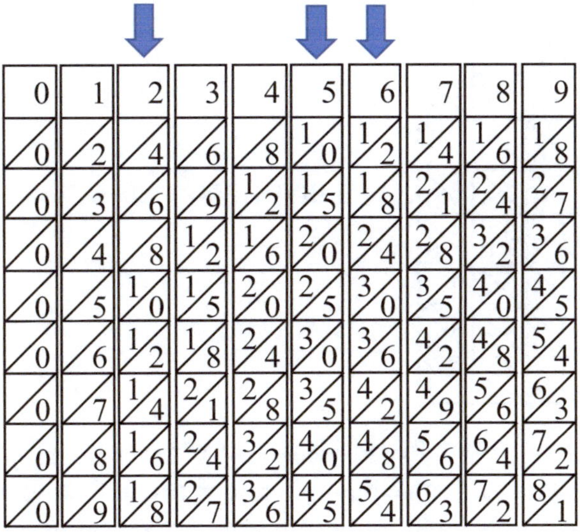

Figure 1.2.9 -3: Tally-x Multiplication for 256 times 45 – Step 1.

Step 2: Place the *multiplicand* sticks next to each other, so it reads correctly across the top, 256. If you have an extra 1s multiplication stick, use it to the left of the multiplicand sticks, as the row *multiplier*. Figure 1.2.9 -4.

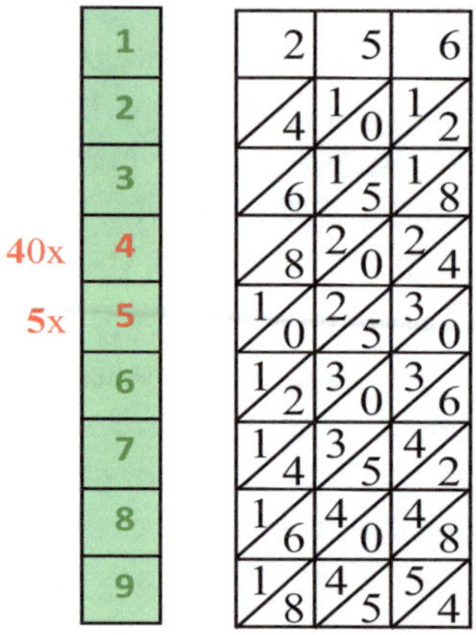

Figure 1.2.9 -4: Tally-x Multiplication for 256 times 45 – Step 2

Step 3: for the 1's place value multiplier, look across the 5th row and write down the numbers from left to right, or across the place values of 1s, 10s, 100s, etc. Values in the same diagonal have the same place value and should be added. If the added answer is greater than 10, then carry the tens value to the next place value to the right. Figure 1.2.9 -5.

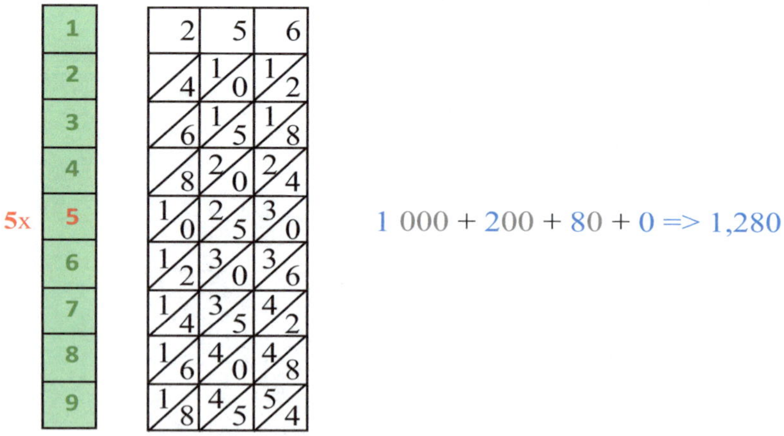

Figure 1.2.9 -5: Tally-x Multiplication for 256 times 45 – Step 3.

Step 4: for the 10's place value multiplier, look across the 4th row and write down the numbers as you did for the 5th row. Remember to carry any value greater than ten. Figure 1.2.9 -6.

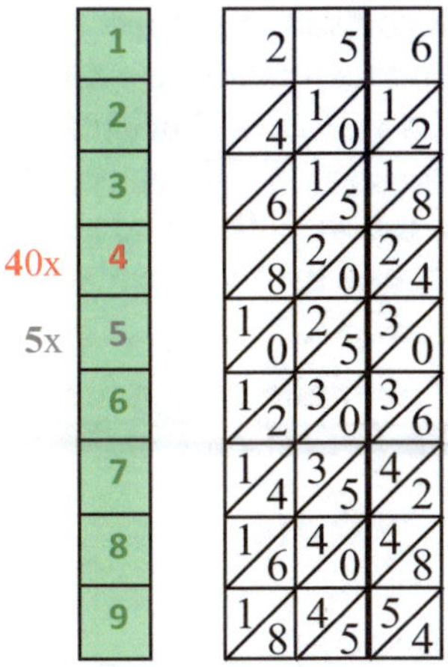

10 000 + 200 + 40 + 0 => 10,240

1 000 + 200 + 80 + 0 => 1,280

Figure 1.2.9 –6: Tally-x Multiplication for 256 times 45 – Step 4.

Step 5: Just as you do for the long multiplication method, you add the 40s answer to the 5s answer. Figure 1.2.9-7.

In steps 3, 4, and 5 it is shown that the values individually for each of the place values show the diagonal addition in the multiplicand sticks. However, that is not necessary. Just write the row answer.

Then, be sure to offset the 40s multiplier to account for it being a 10s place value. The final answer is 11,520 with no memorization or no typo errors. Also, since it is just like the long-multiplication method, you can show your written results to the teacher.

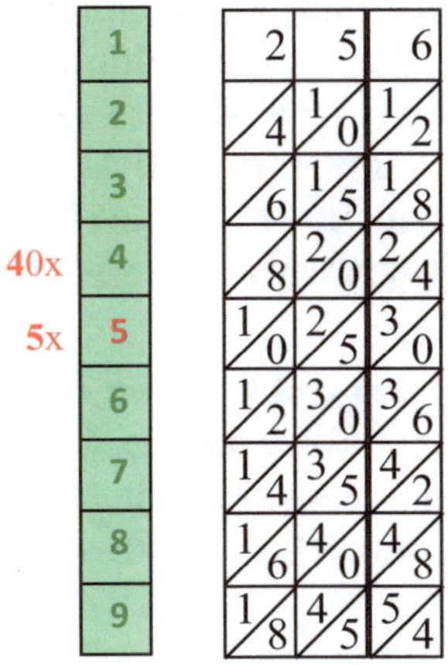

10 000 + 200 + 40 + 0 => 10,240

1 000 + 200 + 80 + 0 => 1,280

11,520

Figure 1.2.9 – 7: Tally-x Multiplication for 256 times 45 – Step 5.

In a second example, all in one step, Figure 1.2.9 - 8 shows the Lattice multiplication method for 525 times 155 = 81,375. In this method, we do it using the place value multiplier to show you a slightly different shortcut. This shows taking the 1 value in the 100s place, 5 value in the 10s place, and the 5 value in the 1s place, the horizontal values add up for 100s place as 52,500, the 10s place is 26,250, the 1s place is 2,625, and then vertically summing those place value results to get the answer: 81,375.

Now it's your turn to try your own multiplicand and multiplier, using the sticks to get the horizontal and the final answer as the vertical sum.

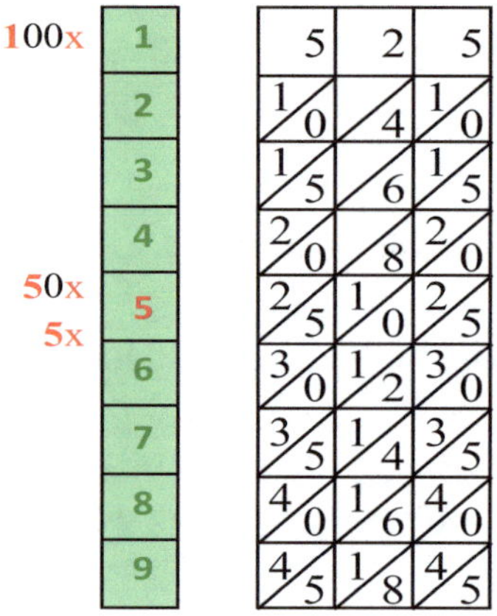

525 x 100 => 52,500

2625 x 10 => 26,250
2625 x 1 => 2,625

81,375

Figure 1.2.9 - 8: Tally-x Multiplication for 525 times 155.

1.2.10 Summary.

Here we have provided some of the key historical points of the multiplication tables from 300 BC up to 1600 AD. Given the effects of the multiplication method in 400 AD with the period for zero, 800 AD with the zero value, 1200 AD with the transition into the Western cultures, and 1600 AD with the conversion from a mathematical method to a calculating tool; one could assume that the next step would be 2000 AD. That is when it was re-discovered again and the source of this book.

This book is founded on the idea left by John Napier and has been expanded to include additional factors: Roadmap of Arithmetic as a pattern, Roadmap to the Curriculums of Math, Learning methods, Patterns of the multiplication table, Multiplication, Division, and Decimal multiplication and division.

1.3: Early Mathematics –A Pattern of Numbers.

There is a pattern in all things of nature and God. Mathematics is all about shortcuts based on empirical observations of patterns in nature. Mathematics dates back to 3,000 BC to the Indus Valley Civilization ("IVC"). All this was covered in the previous sections.

The world contains an infinite variety of patterns, which we discussed. Here we will be looking at the Patterns found in the Multiplication tables themselves. This is why you don't have to memorize the multiplication tables. This book will show you the pattern of these tables and in Chapter 2 recreate them..

If I had you memorize a book on the assembling of a big block motor that puts out 720 horsepower, it would not be impossible but difficult. Most people start with a basic engine to understand the pattern of how it works. Then use this to move up.

In traditional math, you start out memorizing everything you are told. However, will you truly understand how it works? No. If you memorize something, there is a chance you will forget over time. However, if you understand the pattern of something, you will not forget it. That is the Tally-x system of mathematics. We show you the pattern and the process so you can understand and remember more.

Roman Numerals.

In school, the Roman numerals are not introduced to the student until the 7th Grade. It is presented here only to show the progress Western culture gained by switching from the Roman numeral system to the 0-9 system, see Figure 1.3-1.

Roman	I	V	X	L	C	D	M
Digit	1	5	10	50	100	500	1,000

Figure 1.3-1 Roman Numerals.

The original Roman abacus was the place value system of I, X, C, M. But that changed to be more like the hands and its five fingers. This is not a recorded observation, but note that half of ten, the upper half is a V, and so five is V. This is not the case for 50 and 500. There are sources written about this early system of recording values, *reference 1.18*. Today, these Roman numerals are used for fancy clock numbers and are sometimes used in movies to record the year of production. However, the Roman numerals were only useful to record a number, and never to "calculate" a value using the operators of math, i.e., addition, subtraction, multiplication, and division. This is where the abacus was used.

The Roman numerals had a five-spaced Place Value system, i.e., I, V, X, L, C, D, M. The Hindu-Arabic system used a unique Place-value system, where each place-value was limited to the basic 0-9 numbers. This is like an odometer in a car that marks off mileage.

This limitation encouraged Leonardo of Pisa, to accept the Hindu-Arabic system, *reference 1.19*, over the Roman numerals in math. Eventually, Leonardo authored his newly learned mathematical techniques and literally wrote the first math book, *Liber Abaci* ("Book of Calculations"), for the European education system. This was the first dramatic change in the Western world.

0, 1, 2, 3, 4, 5, 6, 7, 8, 9.

These numbers are the way the number sequence should be, based on the help from the Place Value System and the Hindu-Arabic numbers. Not only does it explain the Place Value, but it also eliminates one other issue from Figure 1.3-2.

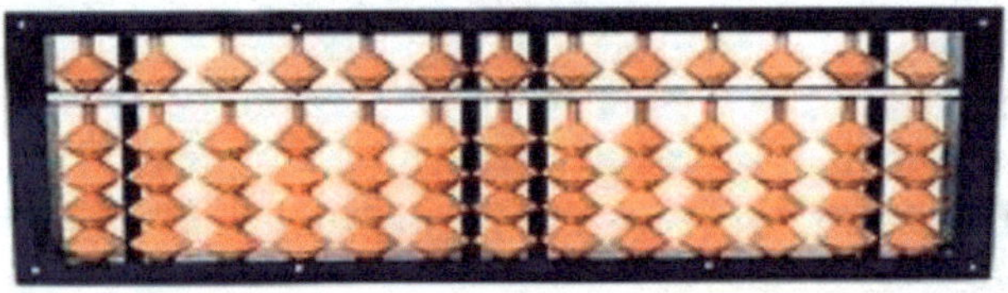

Figure 1.3-2: The Soroban.

With this type of understanding of the Place Value, the student can also understand the most commonly used calculator of all time, the "Abacus." In this case, I am referring to the Japanese Soroban. The Soroban only allows the user to count up to nine on each rod, forcing the 10-digit place value system. Even the Romans understood this method of Place Value by the use of "X" (ten) and "C" (100) symbols. And like the Soroban, the Romans simplified counting by the use of a 5-bead or the "V" (5 count), "L" (50 count), and "D" (500 count) symbols.

This is another example of man's recognizing patterns in numbers. In this case, the V is representative of the number five, but it is also representative of one hand, whereas the X is representative of either both hands or all ten digits. This is seen in Figure 1.3-3, the Chinese abacus Suanpan, where the lower set of beads represents the five fingers on one hand, and the upper set of beads represents the two hands. But, the Suanpan can mistakenly count from 0 to 15 using a hexadecimal system (base-16), and not the decimal system (base 10) of numbers.

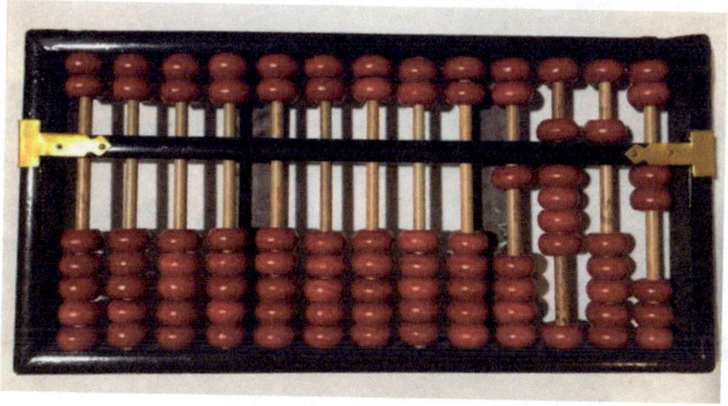

Figure 1.3-3: Suanpan.

If you ever take a first-year software class, you will learn hexadecimal:

0, 1, 2, 3, 4, 5, 6, 7, 8, 9, A, B, C, D, E, F

One of the lessons will be to convert a ten-base decimal addition, subtraction, or multiplication problem into hexadecimal, complete the problem, and finally return it to the decimal value. I did this in my sophomore year in college on a test, in 1976. My teacher inquired why I was playing with the abacus, a toy in his mind, during a test. I explained to him that it was a hexadecimal calculator from China, and I was awarded an A on my test. It was easier to use the Suanpan, by doing the problem on the abacus as hexadecimal and reading it out in the opposite end as hexadecimal or decimal.

1.4: Roadmap of Arithmetic.

Imagine a Life Scenario where every weekday for the next 12 years, you get up each morning and prepare for a new day. You are met by a school bus with other kids your age. This bus has blacked-out windows, so you can't see out. Only the driver knows where you are going and what "he" can see he tells you. Once in a while, you don't go anywhere, and the driver gives you a test on what you remember he told you.

How many years would you endure this, or would you conform to this scenario and do your best?

This is math as it is taught in schools today, which is not a wrong method, as many students do very well with it. However, many do not. Wouldn't it be easier to know where you are going and have a map of what math curriculum you will learn next?

1.4.1 An Alternative to the Life Scenario.

Before we cover the alternative to the Life Scenario, we have to establish a common foundation, what is taught in school and the approach in this book. We will accomplish this by going back to the four basic math operators (e.g., addition, subtraction, multiplication, and division) that are taught in Arithmetic. However, imagine the operators shown in a circle and from the bottom up.

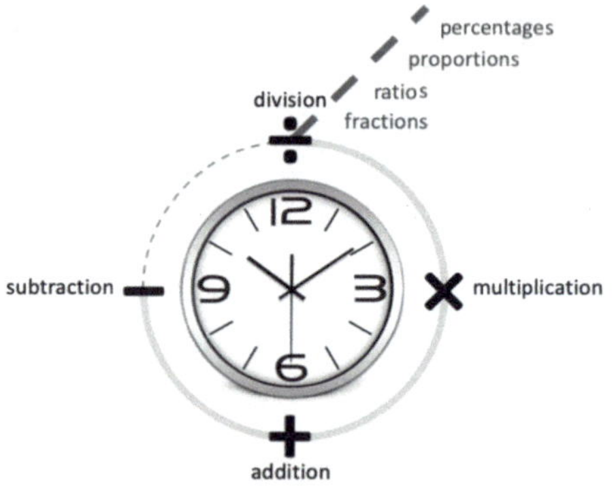

Figure 1.4-1: Roadmap of Arithmetic.

In Figure 1.4-1 is best discussed using a clock's hour position: 3, 6, 9, and 12 o'clock. This is a setup where addition is located at the 6 o'clock position, subtraction is at the 9 o'clock position, multiplication is at the 3 o'clock position, and division is at the 12 o'clock position. With this, let us discuss how these basic operators are sequentially taught and why.

First, Addition, at the 6 o'clock position, is the foundation or prime function and therefore, the first logical approach to arithmetic and all math. The funny thing is that as powerful and fast as a digital computer can be when it comes to math, a computer can only add numbers. Therefore, all mathematical functions in a digital computer are built upon this one operator. In history, addition started with counting pebbles that represented how much you have. The abacus (e.g., Russian -*Tchoty*, Chinese – *Suanpan*, Japanese – *Soroban*, Greek – *Abax*, Aztec – *Nepohualtzitzin*, etc.) is the simple first tool for counting pebbles and can be found around the world.

Addition is built on the numbering system (e.g., 0,1,2,3,4,5,6,7,8,9) where each numeric symbol represents that number of pebbles. The first of the patterns of math or the 10s value system. Take a single pebble and add another one to it and you have two pebbles (i.e., 1 + 1 = 2), where the "+" symbol represents or announces the operational activities, and the "=" symbol equates the answer to the right.

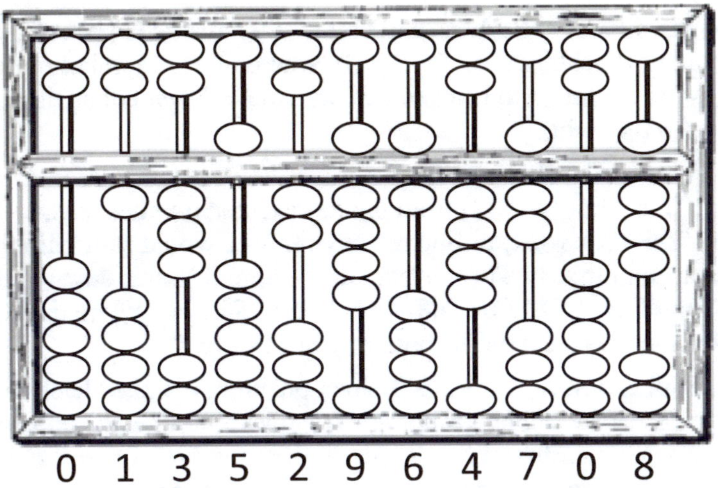

Figure 1.4.1-2: Pebble and Unit Value- Chinese "Suanpan for 1,352,964,708.

To demonstrate addition using the Suanpan abacus (see Figure 1.4.1-2), imagine we want to add four (4) to a number already represented on the abacus, such as 1,352,964,708. Begin by manipulating the beads starting from the 1s column. Move two beads from the bottom row up, which represents a count of five beads, matching the single bead at the top marked for five. Lower this five-count bead and the five individual beads below it. Proceed by moving two more beads from the bottom row up. However, this isn't the end. The two five-count beads at the top equate to a bead in the tens column, the second row from the right. Therefore, move these two five-count beads up and advance one bead in the 10s column. Thus, 1,352,964,708 + 4 = 1,352,964,712.

The addition process with the beads involves moving the remaining two one-count beads from the bottom left to add them to the three beads above, resulting in a total of five beads. This total is equivalent to one five-count bead, causing it to drop down while the five one-count beads descend to the bottom. This accounts for adding two of the original four one-count beads. Finally, raise two of the one-count beads, and your 1s column will display 10 + 2, or 12.

Second, Subtraction, at the 9 o'clock position, this operator is just the *opposite* of Addition, or you are taking away one or more things from a group of things. Interestingly and more of this to come, the area between the "+" operator and the "–" operator is also called the *opposite*. A simple way to remember the differences between addition and subtraction.

So, if you have four (4) pebbles, and you take away one (1) pebble, you end up with three (3) pebbles (i.e., 4 – 1 = 3), where the "–" symbol represents or announces the operation to be executed, and the "=" symbol again equates the answer to the right.

One can even use the abacus, Figure 1.4.1-1, how to demonstrate this operation. To subtract two (2) from the represented number (e.g., 1,352,964,708 – 2 = what) to find the answer, one would start with the one value beads and move two beads from the bottom 3 beads, down to the bottom. Therefore, 1,352,964,708 – 2 = 1,352,964,706. Where the 1 5s bead on top plus the 1 1s bead on the bottom yield 5+1 or 6.

Third, Multiplication "×", at the 3 o'clock position, is linked back to addition as a "shortcut" of adding by grouping the same value. It is the process of adding a grouped number of beads a multiplier number of times. For example, five (5) times three (3) is the addition of three groups of five pebbles. I.e., 5 + 5 + 5 = 15. Or, (1+1+1+1+1) + (1+1+1+1+1) + (1+1+1+1+1) = 15. Therefore 5 times 3 = 15, where the "x" multiplication symbol represents or announces the multiplier operation to be executed, and the "=" symbol again equates the answer to the right.

One can even use the abacus, Figure 1.4.1-2, to demonstrate this operation. However, it is a little tricky because of the hexadecimal numbering capability. However, if I wanted to multiply six (6) times (5), I would have to clear the represented number (e.g., 1,352,964,708) in this figure that no beads are against the center bar (e.g., the counting bar), Figure 1.4.1-3 A.

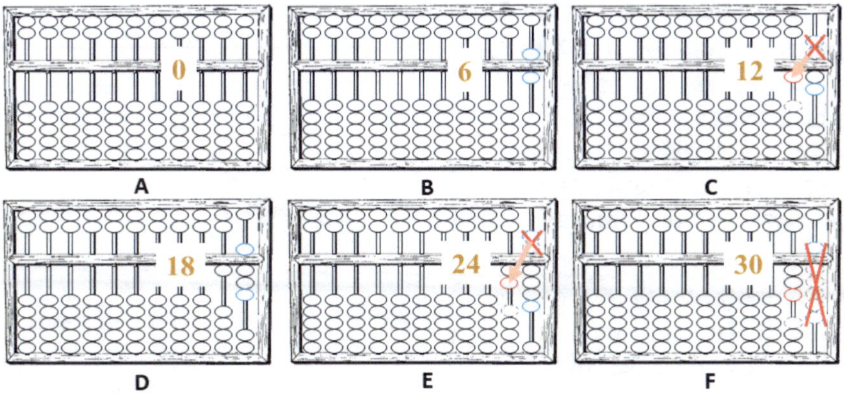

Figure 1.4.1-3: Chinese "Suanpan" Abacus Representing: 6 × 5.

So, the operation is simple, an upper bead (5) and a lower bead (1) representing six (6) are added to the existing bead, Figure 1.4-3 B. This is done five times, were each time being sure to carry to the left as needed, Figure 1.4-3 C, 1.4-3 E, and 1.4-3 F.

Therefore, 6 times 5 = (1+1+1+1+1+1) + (1+1+1+1+1+1) + (1+1+1+1+1+1) + (1+1+1+1+1+1) + (1+1+1+1+1+1) = (6) + (6) + (6) + (6) + (6) = 30.

Fourth, Division "÷", at the 12 o'clock position, is taught as the opposite of Multiplication, just as subtraction is the opposite of addition. So, to know Division, you have to know the multiplication tables. But is that the case?

The purpose of the circle is to show you that:

- Addition and Subtraction are opposites.
- Multiplication and Division are opposites.
- Addition and Multiplication are shortcuts.

So then, Subtraction is a shortcut Division, but in this case, subtracting by a grouping of the same value. Think about it, in a division problem where you are continually subtracting the divisor (left of the box) from the dividend (inside the box). In school, we simplify using multiples of the divisor (multiplication table). However, it could be done by just subtracting the divisor a quotient number of times.

I would recommend getting a Soroban abacus to practice on and learn the basics of bead addition and subtraction, as it is good practice in understanding the place-value system. I would not limit your practicing multiplication and division, but a simpler way is shown in Chapters 4 and 5.

See Chapter 4 on Division for a simple explanation of school-taught long division and the Tally-x version that is compatible with the school approach.

1.4.2 The Extension of the Math Circle.

Zooming in on Figure 1.4.2-1, we see the extension line off of Division, which shows the additional patterns of Division. Consisting of fractions, ratios, proportions, and percentages.

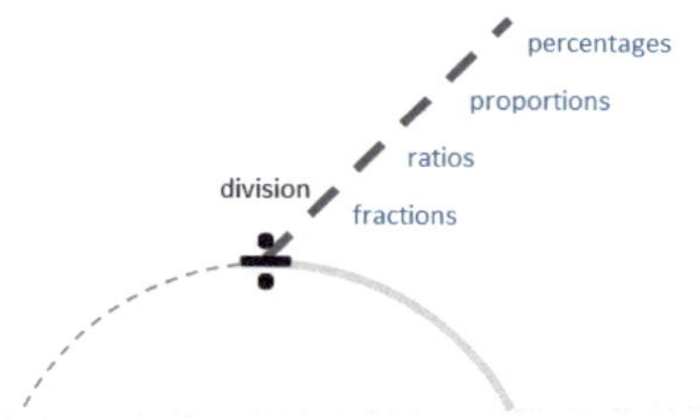

Figure 1.4.2-1: The Division Extension

First, you have the fraction, which is another way of showing division. It also represents an equally represented part of a whole and can also be described as to how many parts of a certain whole. This also references a pie chart and is related to Division, as in Figure 1.4.2-2.

Second, is a *Ratio*, which is a relationship between two numbers, that indicates how many times the first number contains the second. For example, if a bowl of fruit contains three orange slices for every eight grapes, the ratio would be expressed as 3:8.

Figure 1.4.2-2: Pie Chart - shows a Fraction as 3 Slices Out of a Whole Pie of 8- Slices.

Third, is a *Proportion*, where two fractions are equal to each other even though all the numbers are different. They can still be reduced to the same fraction, for example, 3/8 = 9/24. See how this all relates back to the fraction, as the first operator in the division extension.

Fourth, is a *Percentage*, which is a proportion or fraction where one fraction has a denominator of "100". Then the numerator of that fraction is the percentage, for example, 5/20 = 25/100 = 25%. And again, this relates back to Proportion and Fraction.

1.5: Applying the Roadmap to the Curriculums of Mathematics.

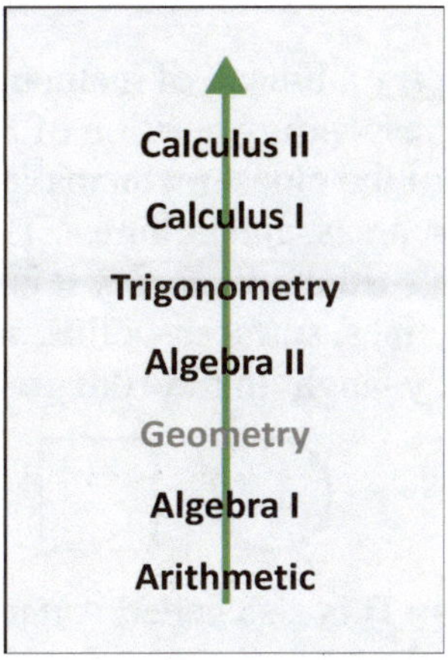

Figure 1.5-1: Curriculum Roadmap of Today's Mathematics.

Figure 1.5-1 is the curriculum roadmap of math, where each of the fundamental operators is repeated in the same way for all courses of mathematics in the first 12 years: Arithmetic, Algebra I, Geometry, Algebra II, Trigonometry, Calculus I, and Calculus II. You will note that Geometery is grayed out, as it is a study of properties of figures: postulates, reasoning and proofs, and definitions.

Understanding the curriculum roadmap helps a student understand and anticipate how new courses in Mathematics (Algebra, Trig, Calculus) are taught. They all follow the same circular Roadmap of the fundamental operators, starting with the *addition of terms*, *subtraction of terms*, *multiplication of terms*, and finally *division of terms*.

The layout is the American K-12 education approach to learning mathematics. However, each state's Department of Education has its own mathematics standards that teachers follow. Also, do not confuse this with the National Council of Teachers of Mathematics (NTCM) – Principles and Standards, the National Assessment of Educational Progress (NAEP) Mathematics assessments, or the U.S. Department of Education.

Arithmetic covers the four operators and is addressed earlier in this chapter. By establishing a solid Arithmetic background in the three basic operations, the following six curriculums are easier, because they are built with the Curriculum Roadmap, Figure 1.5-1.

$1 + 2 = 3 \qquad 3 - 1 = 2 \qquad 2 \times 3 = 6 \qquad 6 \div 3 = 2$

Algebra I sometimes called "Beginning Algebra", refers to the baseline of unknown variables such as times and y, which makes an equation. It is an introduction to the general concepts and terms of algebra. This is where you learn the idea that any equation can be balanced as long as whatever you do to one side of the equation, you do to both. Algebra I is typically taught in the 9th grade.

Another way to think of Algebra I is that in the "beginning," the problems and solutions are simple and you only have to find one or more solutions, based on one or more equations, to a problem.

$$x + 1 = y \quad x + y = 7 \qquad x \times 2 = 2x \qquad x \div 2 = \frac{x}{2}$$

<u>Geometry</u> a branch of mathematics concerned with questions of shape, size, the corresponding position of Figures, and the properties of space. Geometry is one of the oldest mathematical sciences that was initially concerned with lengths, areas, and volumes. These were used to prove the basic operators used in Arithmetic. Today it is concerned with the properties and relations of points, lines, surfaces, solids, and higher dimensional shapes. Geometry is typically taught in the 10th grade.

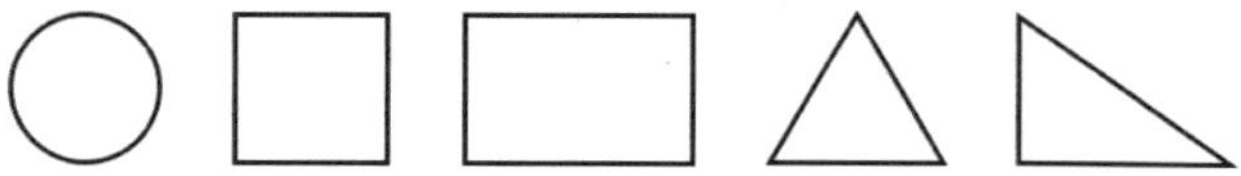

<u>Algebra II</u> is also called "intermediate Algebra," which extends Algebra 1 so that the student can undertake some science courses such as Biology, Chemistry, or Physics. Here you learn about logarithms, complex numbers, functions, and the world of imaginary numbers. Algebra 2 is typically taught in the 11th grade.

Another way to think of Algebra 2 is that the problems and solutions are not as simple, and you may have problems that result in more than two solutions.

Algebra II finds practical application in science and business that includes software, computer engineering, medicine, pharmacist, banking, finance, and insurance.

$$2^2 = 4 \qquad 4^{1/2} = 2 \quad 3^3 = 27 \qquad 27^{\frac{1}{3}} = 3$$

<u>Trigonometry</u> is the study of 2-dimensional Figures such as triangles and how sides are related to angles. Compared to Algebra 2, which is involved in a linear equation. So, in Trigonometry the problems are structured so that there are an infinite number of solutions. Trigonometry is typically taught in the 11th grade.
Trigonometry finds use in the sciences, with applications that include astronomy, navigation, engineering, physics, and geography.

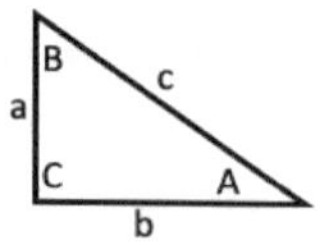

$$\sin\theta = \frac{a}{c,}, \cos\theta = \frac{b}{c,}$$

<u>Calculus I</u> is also called Differential Calculus, where dy divided by dx is the study of change, i.e., speed or distance by time. 'Behavior' here refers to the notion of a 'limit' of a function's behavior at infinity, continuity, differentiability, and integrability. A single variable refers to only one independent variable.

$$f(x) \quad , \quad \frac{d}{dx}f \quad , \quad \frac{dx}{dy} \quad , \quad \frac{dx}{dt} \quad , \int_{c}^{\square} f(x,y)\,ds$$

<u>Calculus II</u> is called Integral Calculus which is the study of "sums" from zero to some variable value, x. An application includes finding the area or volume of non-standard shapes. I.e., waded-up paper, a triangular/rounded eraser, or a soda bottle. It can be thought of as an extension of Calculus I to study functions that are composed of two or more different independent variables.

$$e = \frac{1}{0!} + \frac{1}{1!} + \cdots \quad , \quad \frac{\pi^2}{6} = \frac{1}{1^2} + \frac{1}{2^2} + \frac{1}{3^2} + \frac{1}{4^2} + \cdots$$

Both Calculus I and II are taught in the 12th Grade.

Another way to think of Calculus is that the problems are structured so that the solutions can result in other functions with a more diverse solution and understanding. In each course, we follow the operators in the Roadmap of Arithmetic (e.g., Addition, Subtraction, Multiplication, and Division).

You are merely introduced to something new called "Functions."

In algebra, these functions are the unknowns, or variables, that allow your equation of operators to have one or more solutions to explain something new.

In trigonometry, these functions help explain the relationship of the sides to an angle.

And in calculus, these functions help you understand the relationship (e.g., ratio or proportion) of two or more variables (e.g., non-traditional shapes) or a ratio of a variable with delta time (dt) giving you speed or acceleration

These are principles of the patterns of Mathematics: the patterns of numbers, the pattern of operators, the patterns of functions; in fact, it is about patterns in the learning of math
(Sec. 2.1.4).

Chapter 1 Endnotes.

1.1 2 Kings 20:1, and Isaiah 38:1.

1.2 Past Projects: CV-990 Landing System Research Aircraft (LSRA), 13 August 1993, *https://www.nasa.gov/centers/dryden/history/pastprojects/CV990/index.html.*

1.3 The Soroban is a bead counting tray or abacus developed in Japan. It was derived from the Chinese Suanpan about 1600. The critical difference between the two is one less bead in the heaven (4 counts) and earth (1 count) beads. Thus, one can count to 15 on the Suanpan and only to 9 on the Soroban. Making it faster and easier for 10-digit counting.

1.4 The Abacus is an instrument for performing calculations by sliding counters along rods or in grooves. The word is late Middle English (denoting a board strewn with sand on which to draw figures): from Latin, from Greek *abax*, *abak-* 'slab, drawing board', of Semitic origin; probably related to Hebrew *'āḇāq* 'dust'. But, the first counting tray was the Chinese Suanpan. *http://en.wikipedia.org/wiki/Aba.*

1.5 Carrott, David T, "Tally-x," 2011, ISBN-13 978-0-578-08779-5.

1.6 Indus Valley Civilization was one of the world's first great urban civilizations. It flourished in the vast river plains and adjacent regions in what is now Pakistan and western India (3300 – 1300 BCE). You can find many interesting things about this civilization on the internet, essays, term papers, DVDs, etc.

1.7 The Sanskrit terms Kapâta means "door" and Sandhi means "junctions"; or "the junction of doors"; B. Datta, A.B. Singh, "History of Hindu Mathematics," BKP publication, Delhi India, 2004 edition, ISBN 81-86050-86-8 (set).

1.8 Datta, Bibhutibhushan; Singh, Avadesh Narayan, "History of Hindu Mathematics," Vol 1, pp. 136-137.

1.9 The words are Hindu of Sanskrit origin, where *ganita* means "mathematics" and *manjari* means a "collection" and are from an ancient 16th-century text of Ganesa, the Hindu god of wisdom or prophecy.

1.10 G. A. Grierson (1919). "Augustus Frederic Rudolf Hoernle". The Journal of the Royal Asiatic Society of Great Britain and Ireland. Cambridge University Press. 51: 114–124.

1.11 Corlu, M. Sencer; Burlbaw, Lynn, M.; Capraro, Robert M; Corlu, M. Ali and Han, Sunyoung, The Ottoman Palace School Enderun and the Man with Multiple Talents, Matrakçı Nasuh, p. 23. Journal of the Korea Society of Mathematical Education Series D: Research in Mathematical Education Vol. 14, No. 1, March 2010, 19–31.

1.12 Contact the Bodleian Libraries Communications Office (communications@bodleian.ox.ac.uk) to request a high resolution version of Figures 1.2.4-2 and 1.2.6-2. Credit images per example in the two Figures.

1.13 G. A. Grierson (1919). "Augustus Frederic Rudolf Hoernle". The Journal of the Royal Asiatic Society of Great Britain and Ireland. Cambridge University Press. 51: 114–124.

1.14 Also known as Fibonacci, which is Latin and short for filius Bonacci. It translates into "the son of Bonacci," which was his father's name, Guglielmo Bonaccio.

1.15 Sigler, Laurence, "Fibonacci's Liber Abaci," 2002, Springer Publications, 636 p., ISBN 978-0-387-40737-1.

1.16 John Napier Biography - Famous Mathematicians. *https://www.thoughtco.com/john-napier-biography-4077399.*

1.17. Napier, John, "Rabdology"of 1618, translated by William F. Richardson, Published by the Charles Babbage Institute Reprint Series for the History of Computing, Volume 15, The MIT Press, Cambridge MA and London England, ISBN 0-262-14046-2.

1.18 Hayes, David P., "Guide to Roman Numerals," Georges Ifrah, The Universal History of Numbers: From Prehistory to the Invention of the Computer.

1.19 Hindu-Arabic numerals originated in India in the 6th or 7th century and were introduced to Europe through the writings of Middle Eastern mathematicians, especially al-Khwarizmi and al-Kindi, about the 12th century. Britannica, The Editors of Encyclopaedia. "Hindu-Arabic numerals". Encyclopedia Britannica, 29 Aug. 2022, https://www.britannica.com/topic/Hindu-Arabic-numerals. Accessed 20 September 2022.

Chapter 2: Learning Methods

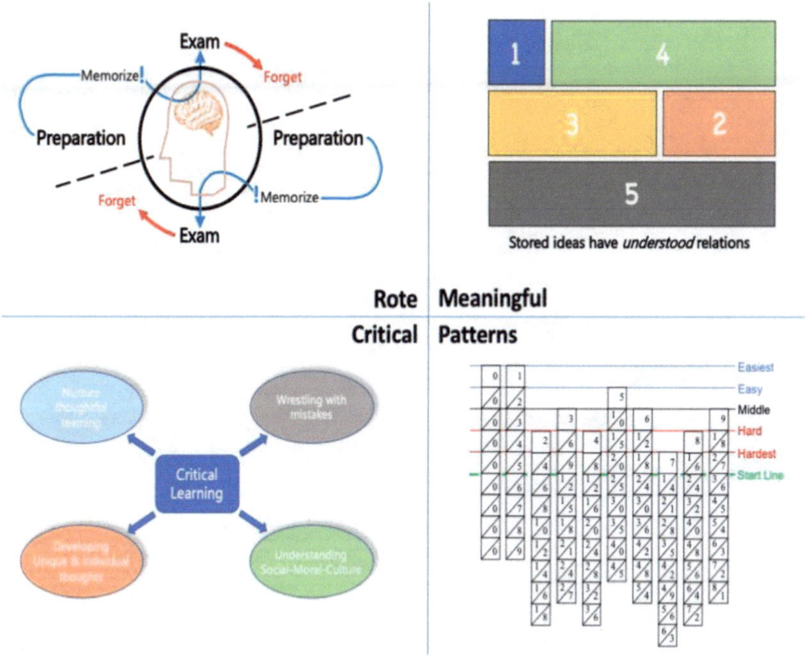

2.1: Learning Methods.

There are many ways of learning: Rote, Meaningful, Critical, Patterns, and others. Each of these methods has a defendable advantage; however, they also come with disadvantages.

If a student memorizes how to assemble an internal combustion engine, it does not necessarily mean they understand how the components interact. Understanding the sequence of which part to assemble before another, does it provide an understanding of their relationship and impact on each other. Questions arise: Do I install the valve cover before assembling the head? Do I install the piston before installing the piston rings? There is a pattern that is sometimes referred to as the art of engine building that works without bringing frustration, and confusion.

Music is an art, yet there is a mathematical pattern in the great classical pieces, and the genre of the types: Classical, Rock, Jazz, Blues, Folk, and more yet to be discovered. Art is art, yet there is also a mathematical pattern to include pictorial representations (works, scenes, views, …) including the Dutch Golden Age, Flemish Baroque, still life, marine, architectural, historical, etc. We can use these mathematical patterns to find copies, i.e., forgeries, of music, art, or fashion designs. It can even be used to find those who commit literary theft, *reference 2.1*.

Let's cover some advantages and disadvantages of these learning methods before we get into this idea of Pattern learning.

2.2: Rote Learning.

The Rote learning method, *reference 2.2*, is the memorization of information based on repetition and is commonly used to defer understanding till later, or not at all. Examples of rote learning include memorizing the alphabet, numbers, and tables. Some consider rote learning to be a necessary step in learning certain subjects such as math, physics, chemistry, biology, etc. Memorization isn't the most effective way to learn however, it's a method many students, and teachers still use. There are many disadvantages to the method of learning: lack of understanding of the subject, no connection between new, and prior knowledge, and wrongful understanding.

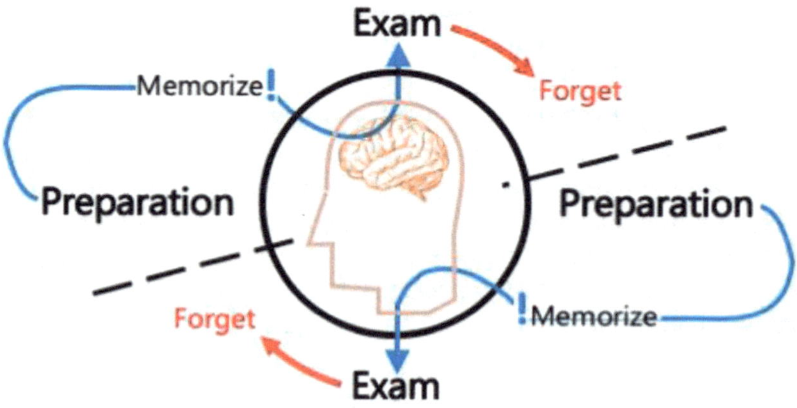

Figure 2.2-1: Rote Learning.

Just before and during World War II, the German political system didn't need people that could think, they needed people who could obey blindly. So, in schools, the students didn't learn how things worked; they memorized things to obey. People like Albert Einstein and many other scientists were treated poorly because they wanted to understand and share their understanding with the world. This was not the way at that time. There have been other historical periods where the general populist was discouraged from understanding and where a few found the Rote approach not satisfying.

When I grew up, we used flashcards to (1) learn to spell words instead of using phonics to understand spelling; (2) learn addition, subtraction, multiplication, and Division; (3) learn history associated with timelines; and even to (4) learn states/capitals of the United States of America. This is the rote system, where knowledge does not necessarily bring understanding or wisdom. In my personal experience, I encountered a gentleman who was considered a subject matter expert in the field of electromagnetism, because he could recite word-for-word and the equations of a book. The trouble was, that he couldn't do the exercises at the end of a chapter unless it required reciting something from the chapter. The problem we were working on required the information derived between two equations. In this case, the second equation was written with the removal of the meaningless terms? However, the work we were doing used "meaningless terms," which this gentleman did not understand.

Everyone with any science study knows about Einstein's famous equation $E=mc^2$. However, did you know that it is only part of the equation? The whole equation is more like the Pythagorean Theorem, *reference 2.3*, of $a^2 + b^2 = c^2$, or where the sum of the square of each of the short sides $a^2 + b^2$ is equal to the hypotonus or the third side (c^2). What is Einstein's full equation?

$$E^2 = (m \times c)^2 + (p \times c)^2 \qquad \text{Equation 2.1.2-1.}$$

Equation 2.1.2-1 states that Energy (E) squared is equal to the sum of the square of the mass (m) and momentum (p). An object with mass cannot move faster than the speed of light because of momentum. Einstein developed and understood the whole equation of why mass cannot go faster than the speed of light. But, not everything has mass, i.e., photons.

Getting back to the Rote system. Isn't there a more natural way to memorize the states, and the capitals? Well, in the late 1990s, a cartoon, *reference 2.4*, on television taught a whole nation all the states and their capitals using music, and the lyrics of a song. An Animaniacs' song included Wakko's 50 State Capitals, to the tune "*Turkey in the Straw*." Why did this work, because the song has a pattern, as does all music, to the words (rhyming), and to the music (rhythm)? This became so popular that other songs were created: Yakko's *World*, Yakko's *Universe*, and Animaniacs' *Presidents*. All based on a pattern of music and lyrics.

In most cases, school tests and exams are set up to reward students who memorize information. Remember the gentleman in my story? Once you've earned a passing grade, there's no need for your brain to remember the information — so out it goes! Memorization is time-limited. But, a pattern is easy to recall and use in many different ways.

Unfortunately, tests like the Scholastic Aptitude Test (SAT), and the American College Testing (ACT) are structured differently. These tests challenge your *knowledge* of real-life and complex ideas to test your *understanding*. Now, as a student, you are faced with a new situation that you didn't previously *memorize*. A pattern can open up these solutions.

2.3: Meaningful Learning.

Meaningful learning involves understanding how all the pieces of an entire concept fit together. With this learning approach, we are getting a little closer to understanding. The knowledge gained through meaningful learning applies to new learning situations. It is active, constructive, and long-lasting; however, it also allows students to be fully engaged in the learning process. They are not repeating a memorized list or set of rules.

Figure 2.3-1: Meaningful Learning - stored ideas have *understood* relations.

The Oxford learning system states that "*Experts emphasize the importance of deep understanding over the recalling of facts. Students who learn with meaningful learning are able to solve problems better than those who learn by Rote.*"

Two important goals of all types of learning include retention and transfer. "Retention" is the ability to remember the material at a later time. "Transfer" is the ability to use prior knowledge to solve new problems.

The key advantages of meaningful learning are: (1) encourages understanding, (2) relates new information to prior information, and (3) relates this new information to future challenges.

2.4: Critical Learning.

Critical learning (or critical thinking) focuses more on engaging with the subject matter, thinking critically, and understanding how all the pieces work together. It helps when students "Learn to think and think to learn." In essence, it encourages understanding rather than memorization. Presented with a math problem, the student uses the criteria to help them choose the most effective strategy.

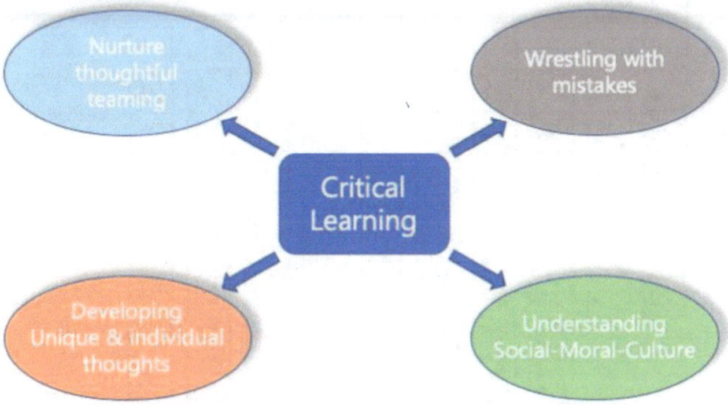

Figure 2.4-1: Critical Learning.

Students use this approach to develop mathematics using:

- Deeper engagement and understanding.
- Greater independence and self-regulation.
- Stronger competence with mathematical processes.

With this type of learning, students "transfer" knowledge rather than just memorizing it — that is, they can take a piece of information, and use it effectively in different contexts and problems. Unfortunately, in most schools, the lessons end before students can reach this level of critical learning. Critical learning may lead to a passing test and getting a good score; however, it doesn't lead to long-term success.

2.5: Pattern Learning.

Pattern learning encourages questions and ideas, and also leads to recognizing patterns of things. When students see these patterns —, and genuinely understand the information at hand — their brain is more likely to retain it.

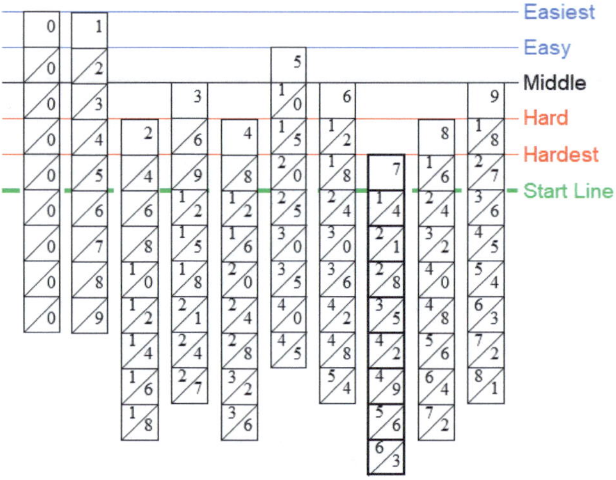

Figure 2.5-1: Pattern Learning.

This method of learning is a collection of the best of the above approaches to learning.

- Easy to recall.
- Developed from foundational knowledge.
- Inter-active / fully-engaged
- Constructive.
- Long-lasting/retention.
- Transferable.

2.5.1 Why do most students 'Hate' Math?

If you ask any student which of the four basic arithmetic operators they "hate," the answer will most likely be Division. Hate is a strong term and sets the general tone of a student's judgment of math. But, there are those few love math. In both cases, this book can be of great value.

The two reasons for these responses are found in the foundation of poor math skills in the United States. Where a student's skill includes a poor understanding of the multiplication tables, and the being taught that division involves 'guessing' the initial answer. i.e., how many times can 6 be divided into 46?

Answer 1 – poor understanding of the multiplication tables.

This is probably the most commonly understood reason why kids 'hate' division. However, the most common response is to have the kids memorize the multiplication table.

During the learning of the multiplication operator, students are expected to memorize the multiplication table for all numbers: 0 through 12, and that is a major request. Some students take to it easily, others have to work at it or work with an adult to help them, and then others never quite get it. I know, I used to be one of the latter examples. Why do we not get it?

There are 12 columns of multiplication to be memorized from 1 to 12. Each column has 12 values. That's 144 values to remember and be able to instantly recall when needed. This is taught in the third grade in the United States when a child is only 9-10 years old. According to the Scholastic Aptitude Test, *reference 2.5*.

> "Understanding the concept behind division can be even harder for kids. As with multiplication, division facts have to be committed to memory with the aid of worksheets, flashcards, songs, and rhymes. Learning to recognize fact families — groups of numbers such as 4, 3, and 12 that create multiplication and division facts in a certain order — can help a child's facility with calculations."

Note the concept of "*committed to memory*" is to memorize something, especially 144 values, is not something easily done. It is the latter part of the statement 'songs and rhymes' that is the key hear. Songs and rhymes are based on a rhythmic or audio pattern, which is one way to remember a pattern. Another way is to use visual patterns and reconstruction. If I show you how to fold paper to create a star that can be hard. However, showing the repeating pattern of folding paper to get an airplane that flies is easily forgotten.

Answer 2 – 'guess'

This is probably the least understood reason why kids 'hate' division. However, if you think it through, it makes sense. Mathematics is an exact science. Wikipedia says that "*an exact science is any field of science capable of accurate quantitative expression or precise predictions, and rigorous methods of testing hypotheses, especially reproducible experiments involving quantifiable predictions and measurements*".

Let's take a moment and look at what we mean by mathematics.

Mathematics is the study of quantity, space, structure, and change. Mathematicians seek out patterns and formulate new conjectures. Mathematicians resolve the truth or falsity of conjectures by mathematical proofs, which are arguments sufficient to convince other mathematicians of their validity.

So, it is that mathematics is so exact that it is used universally across languages and nationalities. It is the most simplistic form of communication and understanding. For example, 2 + 4 is always 6 when using the addition operator, and 2 times 4 is always 8 when using the multiplication operator. Similarly, 8 divided by 4 is always 2 when dividing. There is no other solution to these values.

The problem is that to make a 'guess' implies a chance of failure. You get it right, or you get it wrong. However, in either case, you have to know the answer, and as such, you have to know the multiplication tables. In the above-cited example of division, 8 ÷4, one could guess 1 or 3. In the first case, the answer would be 2, which is too low. In the second case, the answer would be 2.666, which is too high. So, you try 2, which is 8, and just right. However, what if the problem was 10 divided by 4? Knowing the multiplication table will never reveal the correct first answer. 1, 2, and 3 will never yield 10.

This process of guessing is an excellent introduction to logic, except for most students just learning the four basic arithmetic operators, it is too soon. Hence, the confusion.

2.6 Are There Organizations Against Math Memorization?

The funny thing is that there are organizations that disagree with Rote learning. The National Council of Teachers of Mathematics (NCTM) was founded in 1920. It has grown to be the world's largest organization concerned with mathematics education, serving members throughout the United States and Canada.

> "More than ever, mathematics must include the mastery of concepts instead of mere memorization and the following of procedures. More than ever, school mathematics must include an understanding of how to use technology to arrive meaningfully at solutions to problems instead of endless attention to increasingly outdated computational tedium."
> *reference 2.6.*

Rote learning is strongly discouraged by many new curriculums of math standards. For example, science, and mathematics standards in the United States specifically emphasize the importance of deep understanding, over the mere recall of facts, which is seen to be less important. But, at present, the Programme for International Student Assessment (PISA) results, from 2015, placed the U.S. an unimpressive 38th out of 71 countries in math, 24th in science, and the 35 members of the Organization for Economic Cooperation, and Development, which sponsors the PISA initiative, ranked the U.S. at 30th in math. How do we fix our broken system?

Chapter 2 Endnotes.

2.1 Merriam-Webster states in their online dictionary, that plagiarism means (1) to steal, and pass off (the ideas or words of another) as one's own, (2) to use (another's production) without crediting the source, (3) to commit literary theft, or (4) to present as new, and original idea or product derived from an existing source.

2.2 "The Difference between Rote Learning, and Meaningful Learning," Oxford Learning, 23 March 2017, *www.oxfordlearning.com/difference-rote-learning-meaningful-learning*.

2.3 Named for Pythagoras (Greek), but was well known before Pythagoras, but history states he was likely the first to prove it using geometric rectangles.

2.4 Animaniacs (1993-1998), "Amazing Animaniacs Songs that Taught Us more Than School": Presidents, Nations of the World, Wakko's America (State/Capitals), Yakko's Universe, The Senses, Multiplication Song.

2.5 *www.scholastic.com/resources/article/times-tables-and-more/*.

2.6 On April 12, 2000, The National Council of Teachers of Mathematics (NCTM) released Principles, and Standards for School Mathematics (PSSM), a 402-page revision of the NCTM Standards.

Chapter 3: Patterns of the Multiplication Table

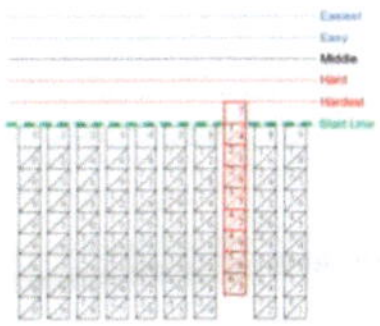

1

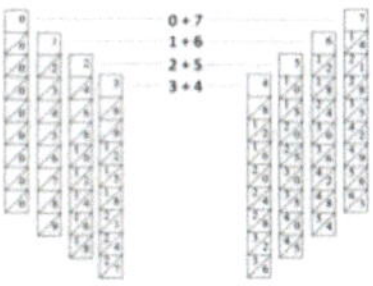

2

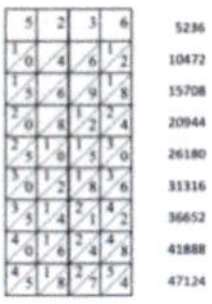

3

4

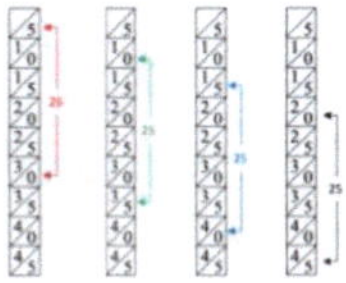

5

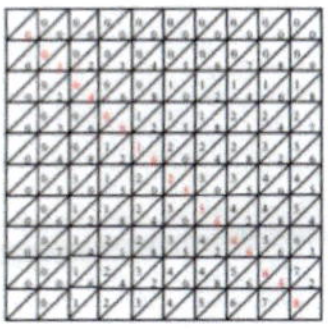

6

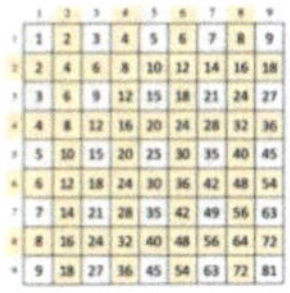

7

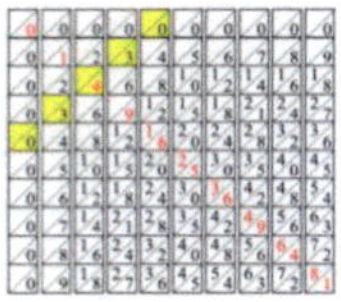

8

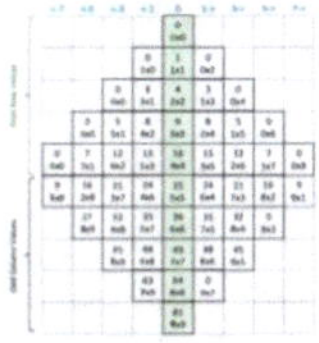

9

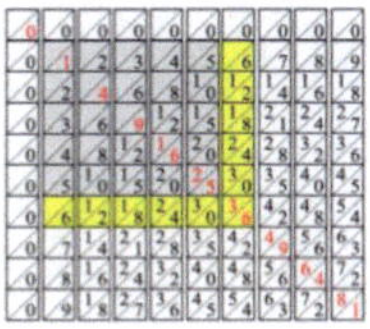

10

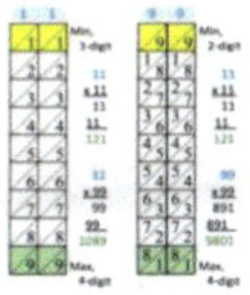

11

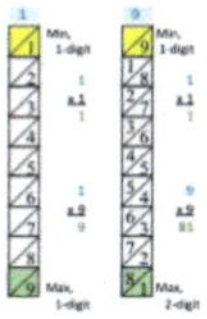

12

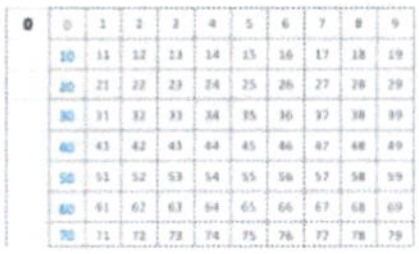

13

32 => 3+2 => 5 ⎫
+ 11 => 1+1 => 2 ⎭ => 5+2 => (7)
43 => 4+3 => (7)

14

15 Digital Root: 64 => 6+4 = 10 => 1+0 => 1

16 Tesla's Digital Root Circle

17 Odd and Even Patterns

3.0: Patterns of the Multiplication Table.

There is a pattern in all things of nature and God. The Lord said, "Mine house is a house of order …, and not a house of confusion." If we pattern our homes according to God's guidance, we will be better. Even to those of the Muslim faith, there is a pattern, or chronological order, to the suras. The order in which the Quran surahs are revealed to the holy prophet, and to guide man. Most religions have a pattern to their belief(s) and as such an order to understanding. Even the natural world contains an infinite variety of patterns. Patterns are found in plants, foliage, and animals. All living things create patterns. Even the Earth with its seasons and the rotation of the stars have an order or pattern. There are patterns in the sand dunes created by blowing winds. There is a pattern in the vortex of a whirlpool, and the formation of an ice crystal. As such, these can all be defined in mathematics. Patterns are also continually being created by simple physical laws.

In mathematics, there is a science of Chaos called Fractals. At first, it seems counter-intuitive that math with its patterns could be used to explain "Chaos"; however, it is the focus on the "behavior" of a dynamic system based on an intricate or hinged pattern. Fractal science is a fascinating mathematical/graphic field that gives us a new way to think about the natural world. Benoit B. Mandelbrot coined the term "fractal" in the 1960s. He spent his life shocking the mathematical establishment and is revered for inventing a new geometry.

An excellent example of a pattern of dynamic is a simple double-rod pendulum, Figure 3.0-1.

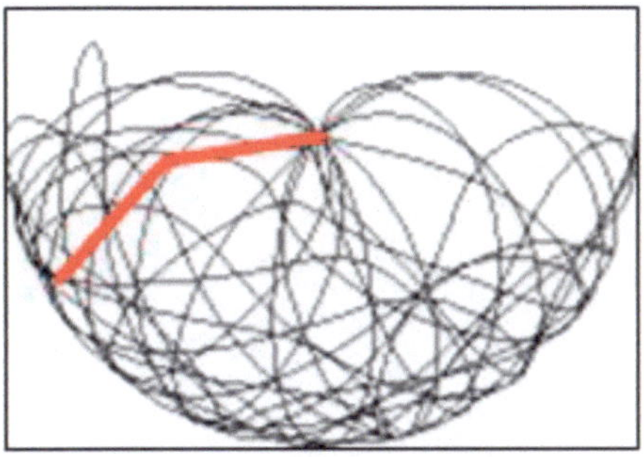

Figure 3.0-1: Double-rod Pendulum.

A double-rod pendulum animation demonstrates a chaotic behavior. Starting the pendulum from a slightly different initial condition, one of the three chaotic properties of Robert L. Devaney, *reference 3.2*, would result in a completely different trajectory.

However, even simple mathematics such as Arithmetic operators (addition, subtraction, multiplication, and division) with their affected numbers have a pattern, which is what this section will show you.

3.1 Geometrically Understanding Multiplication or Area.

Let us start with the basic multiplication table. In Chapter 1.1.3, we discussed the development of the 10-digit numbering system using rows and columns, thus showing the place value system. This ended with Figure 1.1.3-2, the proper listing of numbers. We now moved on to the Soroban abacus, Figure 1.1.3-3. Here we start with the same idea row, and column idea of Chapter 1.1.3; however, the blocks are blanks.

	0	1	2	3	4	5	6	7	8	9
0										
1										
2										
3										
4										
5										
6										
7										
8										
9										

Figure 3.1-1: Proper Column (multiplicand) and Row (multiplier) Table.

In Figure 3.1-1, the blank cells are to be filled in with the multiplication product of each column (multiplicand) by row (multiplier). This is the basis of the Rote learning system.

3.2 The Multiplication Table: 0 to 9.

In school, especially in the United States, students are told to memorize the first 12 multiplication tables. See Figure 3.2-1. There are two flaws to this teaching method, which dates back to the early 1920s. The first flaw is that when a student learns this system, they assume then that they must also memorize in some future time the rest of the multiplication tables, 13 and up. This is one of the reasons why most students in their early years of math, learn not to like it. It is too hard.

The second flaw is that most teachers and teachers of teachers don't know why we require the memorization of the 12 multiplication tables. I have asked this question of many education professors and found many answers. However, they are all guess work, because they don't know or say it is how it has always been done.

	0	1	2	3	4	5	6	7	8	9	10	11	12
0	0	0	0	0	0	0	0	0	0	0	0	0	0
1	0	1	2	3	4	5	6	7	8	9	10	11	12
2	0	2	4	6	8	10	12	14	16	18	20	22	24
3	0	3	6	9	12	15	18	21	24	27	30	33	36
4	0	4	8	12	16	20	24	28	32	36	40	44	48
5	0	5	10	15	20	25	30	35	40	45	50	55	60
6	0	6	12	18	24	30	36	42	48	54	60	66	72
7	0	7	14	21	28	35	42	49	56	63	70	77	84
8	0	8	16	24	32	40	48	56	64	72	80	88	96
9	0	9	18	27	36	45	54	63	72	81	90	99	108
10	0	10	20	30	40	50	60	70	80	90	100	110	120
11	0	11	22	33	44	55	66	77	88	99	110	121	132
12	0	12	24	36	48	60	72	84	96	108	120	132	144

Figure 3.2-1: 12-value Column - Multiplication Table.

In the 20 years of asking "why 12" only two professors that teach math to math teachers provided a different answer: (1) there are 12 inches to the foot, or (2) there are 12 months in a year. Neither of these answers makes any sense in the association with the multiplication table. This again is another contribution to the student's confusion, and lack of interest in math.

Just looking at this grid of 144 numbers (12x12) can be intimidating to any grade-schooler. Imagine the "joy" of memorizing each and every cell. The idea is not based on a 10s Place value system of numbers, and it would appear at first that memorization of this 12x12 grid was just the beginning. Would the student be required to memorize 13x13 or an even larger grid of 50x50 (2,500 numbers)?

However, to add to the confusion, the school approach is to throw out this 12x12 grid when introducing the long multiplication method. This is a 10-place value system using single-digit multipliers. Why then did the student have to learn a 12x12 multiplication table? This is the heart of the memorization approach and the heart of introducing multiplication in grade school. Can't there be a more straightforward approach to understanding?

3.3 10s Value Multiplication Table.

If we go back to Figure 3.3-1, which is a 10x10 grid (0 to 9), we can fill in the product in each blank square, by multiplying the column value (multiplicand) by the appropriate row valve (multiplier), see Figure 3.3-1.

	0	1	2	3	4	5	6	7	8	9
0	0	0	0	0	0	0	0	0	0	0
1	0	1	2	3	4	5	6	7	8	9
2	0	2	4	6	8	10	12	14	16	18
3	0	3	6	9	12	15	18	21	24	27
4	0	4	8	12	16	20	24	28	32	36
5	0	5	10	15	20	25	30	35	40	45
6	0	6	12	18	24	30	36	42	48	54
7	0	7	14	21	28	35	42	49	56	63
8	0	8	16	24	32	40	48	56	64	72
9	0	9	18	27	36	45	54	63	72	81

Figure 3.3-1 10s value Multiplication Table.

Although this can be a very practical approach to multiplication, it can be improved using John Napier's (Chapter 1.2.9) set of rods that put the one value in the lower part of the square, and the tens value in the upper part.

Organizing the Multiplication Table Using 10s, and 1s place value arrangement provides an approach to doing carrying during multiplication. It also provides a clearer understanding of the place value meaning in the number.

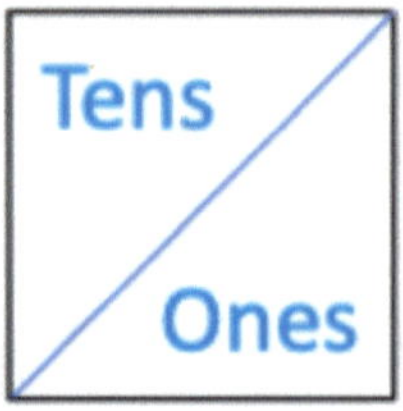

Figure 3.3-2: Arrangement of the 10s, and 1s Place Value in Each Square.

Using this approach, the 10s value multiplication table in Figure 3.3-1 now becomes the assemblage of the multiplication sticks, and is shown in Figure 3.3-3.

	0	1	2	3	4	5	6	7	8	9
0	/0	/0	/0	/0	/0	/0	/0	/0	/0	/0
1	/0	/1	/2	/3	/4	/5	/6	/7	/8	/9
2	/0	/2	/4	/6	/8	1/0	1/2	1/4	1/6	1/8
3	/0	/3	/6	/9	1/2	1/5	1/8	2/1	2/4	2/7
4	/0	/4	/8	1/2	1/6	2/0	2/4	2/8	3/2	3/6
5	/0	/5	1/0	1/5	2/0	2/5	3/0	3/5	4/0	4/5
6	/0	/6	1/2	1/8	2/4	3/0	3/6	4/2	4/8	5/4
7	/0	/7	1/4	2/1	2/8	3/5	4/2	4/9	5/6	6/3
8	/0	/8	1/6	2/4	3/2	4/0	4/8	5/6	6/4	7/2
9	/0	/9	1/8	2/7	3/6	4/5	5/4	6/3	7/2	8/1

Figure 3.3-3: 10s Multiplication Table in the Form of Vertical Sticks.

3.4 Patterns.

The following patterns of 1 through 9, can be used to use the single-digit multiplication table to understand the tables' multiple patterns of understanding multiplication and division, and the functional properties that include the cumulative Properties.

For ease of learning, the Patterns are grouped into the *Table-based*, the *Column-based*, and the *Digit-based*.

The *Column-based* group is all about the multiplication table as columns of multiplication sticks, see Figure 3.4-1, and their relationship to each other. So, in this case, it is about the individual parts of the Multiplication Table, whereas the Table is covered in the next Group.

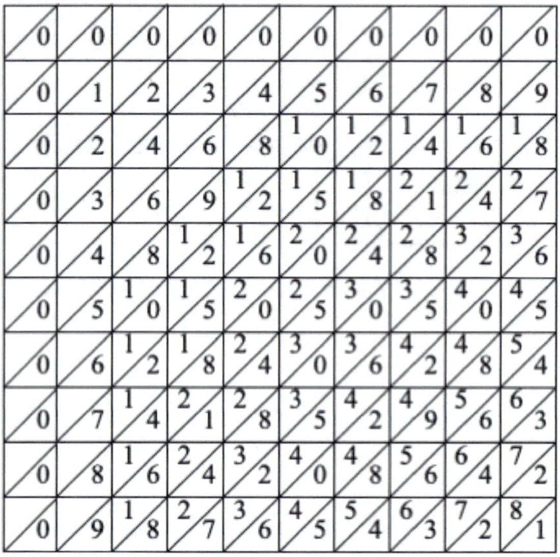

Figure 3.4-1: Column Based Form.

The first five (5) patterns 1, 2, 3, 4, and 5 are best explained if one follows along using the Tally-x sticks, see Chapter 3.5 to copy or create your own.

The *Table-based* group is about the multiplication table as a whole. It is the relationship of the columns, rows, and diagonals in the next five (5) patterns of this grouping; Patterns 6, 7, 8, 9, and 10. In this case, the patterns are seen as mirroring or complementary diagonal relationships of numbers.

In all the patterns of this grouping, everything is seen starting with the multiplication table as a whole, see Figure 3.4-2.

	0	1	2	3	4	5	6	7	8	9
0	0	0	0	0	0	0	0	0	0	0
1	0	1	2	3	4	5	6	7	8	9
2	0	2	4	6	8	10	12	14	16	18
3	0	3	6	9	12	15	18	21	24	27
4	0	4	8	12	16	20	24	28	32	36
5	0	5	10	15	20	25	30	35	40	45
6	0	6	12	18	24	30	36	42	48	54
7	0	7	14	21	28	35	42	49	56	63
8	0	8	16	24	32	40	48	56	64	72
9	0	9	18	27	36	45	54	63	72	81

Figure 3.4-2: Table Based form.

The *Digit-based* group has nothing to do with the columns or the table as a whole and is something that is seldom taught in school. It is the heart of why the place values system of the Hindu-Arabic numbers works. Pattern 11 can be seen by example using the Tally-x sticks or doing long multiplication. It was first developed for division because of the way division is taught – Guessing how many times a divisor can go in into a dividend. This one Pattern can help anyone to understand division better.

Of these patterns, the Tally-x method work is best learned with Pattern 1: Easy to Hard, Pattern 2: Addition of Multiplication, Pattern 3: Multi-digit Multiplicand numbers, and Pattern 12: Place value.

All the other Patterns (4, 5, 6, 7, 8, 9, 10, 11, 13, 14, 15, 16, and 17) reinforce the idea that there are many ways to understand the multiplication table. This insight is not necessary in the 2nd or 3rd grade, but should be learned before pre-algebra is studied.

3.4.1: Pattern 1 – Easy to Hard Multiplication.

In this pattern, the approach is to organize the individual columns of the multiplication table from easiest to more difficult patterns of multiplication columns. This is not based on prior memorization unless you already have.

3.4.1.1 Setup.

First, make a copy of the multiplication table, and cut the columns into strips. See Figure 3.4-1 as an example. Layout out each column on a table or tape them to a whiteboard.

Figure 3.4.1-1: Easy to Hard – Start Position.

Typically, you would be required to memorize these plus ten, eleven, and twelve. We'll address ten, eleven, and twelve in a future pattern. So, for now, we are going to move them up according to which are easy or hard so that we can see a pattern in these multiplication tables.

3.4.1.2 The Multiplication Column of **Zero** (0).

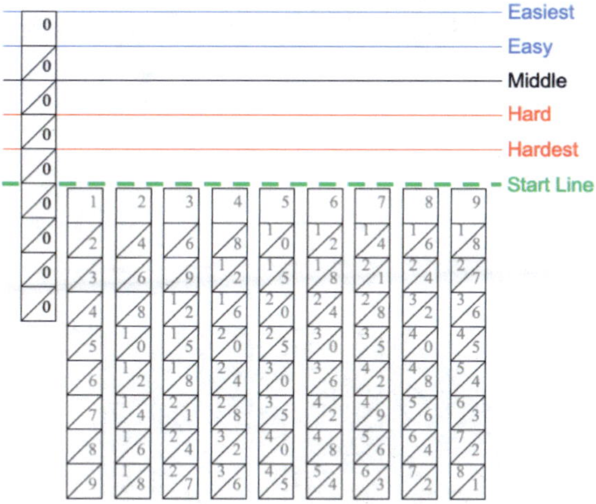

Figure 3.4.1-2: Easy to Hard – Zero (0).

Zero: is the easiest multiplication table because nothing times anything is nothing. The most straightforward pattern of multiplication is any number multiplied by zero is 0. See the Zero (0) column in the above figure. This is the Properties of Zero for multiplication.

The 0s multiplication table is the easiest to understand, because in the multiplication Properties of Zero: any number multiplied by zero is zero. It means that if you have no values of any number, you have nothing. If you have zero sets of any number, you have nothing. However, if your set is empty, { }, then you also have nothing. I.e., 0x5 =0. This is normally taught in the 5th grade, that anything time zero is zero.

But, history has been hard on this simple value, zero. It was the use of place values, where 10s, 20s, 60s, or any other that zero became important.

The introduction of a missing position or zero possibly occurred in 876 A.D. in India or by Hindus with the decimal system, or even in Alexandria as a Greek value. The Hindus notation for this tenth value was a round "*goose egg*". Religions separated the Western world from the Hindu and each developed their number values opposingly. So, it wasn't until the 1200s AD under Fibonacci, see Section 1.2.8, that the Hindu-Arabic numbering system was introduced into the European or Western world. Then only through his book, Liber Abaci were the full number values (0-9) introduced because it allowed for non-abacus methods of calculating the complexities of math such as multiplication and division. Unfortunately, it did not come into general use until the 16th century. To complicate this zero even more, the symbol for zero

"0" or *goose egg*, went in and out of favor depending on the authoring Western mathematician. For example, the slanted solidus "/" was suggested in 1845 by Augustus De Morgan University College London.

3.4.1.3 The Multiplication Column of One (1).

Figure 3.4.1-3: Easy to Hard – One (1).

One: is still easy because the answer is the multiplicand, as the answer. From the first day, we learn to count, we know the multiplication of ones. Each value after one (1) is just one more than the last. The pattern is simply counting to nine. Therefore, the multiplication Properties of one: any number multiplied by 1 is that number. I.e., 1x4 =4. This, again, is normally taught in the 5th grade, that any number times one is that same number, including across the place value system. For example: 1,369 x1 = 1,369.

Note: The 1s table is commonly used with Tally-x to provide the multiplier value for each column. i.e., the row is the multiplier.

3.4.1.4 The Multiplication Column of **Five** (5).

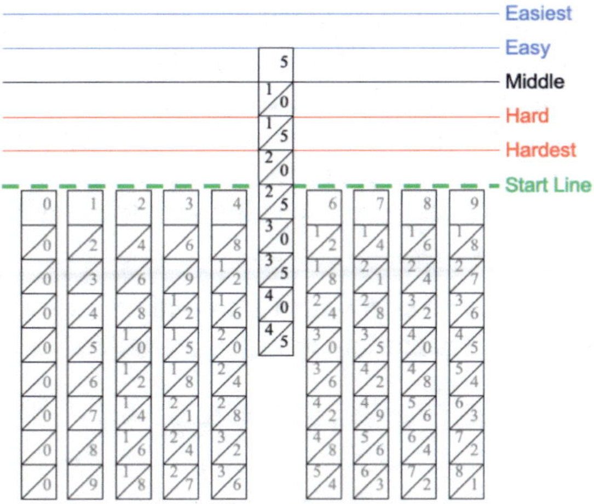

Figure 3.4.1-4: Easy to Hard – Five (5).

Five: is like the Zero (0) and One (1) multiplication column, in that it is unique to itself. However, unlike the other two, it has a rhythm that you might be familiar with in rhyme or a cheerleading shout-out: two, four, six, eight, who do we appreciate. But, this only applies to four of the five numbers in the 5s stick. Missing would be zero and all the two-digit values of 10, 12, 14, 16, 18.

One way to look at the pattern of the 5-stick involves the 1s and 10s place values:

> 1s place value: is a repeating pattern of 0, 5, 0, 5, 0, 5, … 0, 5, … . This is infinity repeating. Note: that any number ending in 5 is a 5-multiplier. Whereas any number ending in 0 can be a 5- or 10- multiplier.
>
> 10s place value: is based on the idea that it takes two nickels to make a dime. Therefore, the 10s place value repeats, i.e., 0, 0, 1, 1, 2, 2, 3, 3, .. 8, 8, 9, 9.

If you have an understanding of odd and even numbers, then there are two more patterns in the 5s multiplication table. This is associated with the odd values (1, 3, 5, 7, 9) and the even values (0, 2, 4, 6, 8) as multipliers.

1) For the 1s place value:
   a) If the multiplier is even then 1s place value will be zero. i.e., 2 x5 = 10.
   b) If the multiplier is odd, then the 1s place value is five *reference 3.3*. i.e., 3 x5 = 15.
2) For the 10s place value:
   a) If the multiplier is even then the 10s place value will be half the multiplier. i.e., 4 x5 = 20. Note: half of 4 is 2 and the 1s value is 0.

b) If the multiplier is odd, then the 10s place value will be half the multiplier of the even value below it.

Most people are satisfied with the knowledge that the 1s value alternates between 0 and 5. I.e., 0, 5, 0, 5, 0, 5, 0, 5. But, this is knowledge without understanding. Adding the fact that the 10s place values are also repeating twice for every value from 0-9; that is knowledge or understanding of the pattern of the 5- stick. I.e., 00, 05, 10, 15, … 40, 45. It is better to understand the relationship of the 10s and 1s place value with the multiplier.

The following 2s patterns all start with a single-digit stick but include its cousin patterns of 4 and 8 according to Egyptian understanding.

3.4.1.5 The Multiplication Columns of **2, 4, 6, 8**

Figure 3.4.1-5: Easy to Hard – 2,4,8.

Two: The pattern of 2s is a matter of doubling the pattern of 1s. The student can also learn the rhythm of 2s in counting, starting with the two: 2, 4, 6, 8 for each value of the 1s place value, plus zero (0). It then repeats with the first 10s value, and again, and again. A very simple pattern often used in football cheers, "Two, Four, Six, Eight - who do we appreciate."

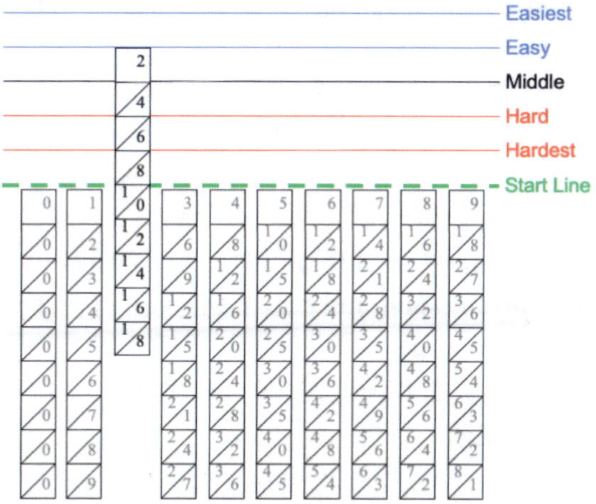

Figure 3.4.1-6: Easy to Hard – two (2).

The unique feature of this is that these numbers (2, 4, 6, and 8) are all multiples of two, even numbers, and therefore included in the two's multiplication table.

2, 4, 6, 8, 10, 12, 14, 16, 18

However, what if we continue this sequence using the Digital Root? What is a Digital Root? Well, if your number has more than one digit, the digital root is the sum of those digits, or the repeated summing of those digits until a single digit is obtained. For example, in Table 3.4.1-1, you have the last six values of the 2s multiplication:

10 ⇨ 1+0 = 1   12 ⇨ 1+2 = 3
14 ⇨ 1+4 = 5   16 ⇨ 1+6 = 7
18 ⇨ 1+8 = 9

Table 3.4.1-1: 2's Stick Digital Root.

Notice that in the first set of 10s values of 0, you get even 1s values that are in the rhyme 2, 4, 6, 8. This is because of the *Even+Even* Rule,

Adding a 0 or even number to an even number yields an even number.

But, in the second set of 10s value you have 1, Table 3.4.1-1, and you get an odd number. This is because of the *Odd+Even* Rule,

Adding an odd number to an even number yields an odd number.

Do we need to continue through 100 values? No, because in the next set of 20s and 30s, the above pattern repeats. Why? Because the of above two rules, the sequence repeats for all future even and odd 10s values, such as the 20s and 30s, 40s and 50s, 60s and 70s, etc.

Four: is as easy as the twos stick because it is just doubling (2 times) the 2s values.

Doubling was one of the earliest patterns noted by the Egyptians.

For example: 2x2 is 4, 4x2 is 8, 6x2 is 12, 8x2 is 16, etc.

However, the Egyptians also noticed that it meant the second value in the 2s column, or skipped every other value. So, given the 2s column, to the right, the second value is 4, then the second jump is 8, then 12, then 16, etc.

Now, let's look at 2s double or 4 and repeat the same two rules above for digit-sum. You get:

4, 8, 12, 16, 20, 24, 28, 32, 36

But, the digital root yields:

4, 8, 3, 7, 2, 6, 1, 5, 9

This is all the numbers 1 to 9 but in a rearranged order. This arrangement can be seen using the Egyptian approach to doubling of 2. Instead of simply multiplying 2 to get the 4s-stick, the Egyptians viewed it as a pattern derived from the 2s-stick. They observed that the sequence of the 4s-stick resembles a shifted version of the sequence of the 2s-stick: 2, 4, 6, 8, 12, 14, 16, 18. Doubling starts from the number two, and you look at every second value of the sequence: 2, 4, 6, 8, 12, 14, 16, 18. The subsequent value in the 4s sequence involves looking at the next second value, as shown below:

2, **4**, 6, **8**, 10, **12**, 14, **16**, 18, **20**, 22, **24**, 26, **28**, 30, **32**, 34, **36**, 38

Notice that this sequence also results in digital roots of 4, 8, 3, 7, 2, 6, 1, 5, 9. Note that there are 2 even values, followed by 2 odd values, then 2 even values, and finally 2 odd values, plus the number nine.

Six: is as easy as the 2s stick because it is just triple (3 times) the 2s values.

6, 12, 18,   24, 30, 36,   42, 48, 54

But, the digital root yields:

6, 3, 9,   6, 3, 9,   6, 3, 9

Tripling was also one of the earliest patterns noted by the Egyptians.

For example: 2x3 is 6, 4x3 is 12, 6x3 is 18, 8x3 is 24.

However, the Egyptians also noticed that it meant the second value in the 2s column, or skipped every other value. Given the 2s column, also to the right, the third value is 6, then the second jump is 12, then 18, etc.

This is also covered, in a different way in the next 3, 6, 9 sequences, which may be even easier.

Eight: is as easy as 2s triple or 8, using the same 1 through 9 multipliers. In this case, you get:

8, 16, 24, 32, 40, 48, 56, 64, 72

But, again the Egyptians understood it as the triple skipping of the 2s stick. They noticed that if you look at the 8s stick sequence above, you get the skipping of every third value in the 2s stick.

2, 4, 6, **8**, 10, 12, 14, **16**, 18, 20, 22, **24**, 26, 28, 30, **32**, 34, 36, 38, **40**, 42, 44, 46, **48**, 50, 52, 54, **56**, 58, 60, 62, **64**, 66, 68, 70, **72**

Notice, that there are 3-gray values in front of each 4s value, which is the Egyptian understanding. And if we apply digital root to the highlighted values we get: 8, 7, 6, 5, 4, 3, 2, 1, 9; or the values 1-9 arranged in a descending order from 8 to 9.

In summary, I would link 2, 4, 8 as a sequence of Egyptian doubles and triples of the 2s stick and leave the 6s sequence for the next pattern.

3.4.1.6 The Multiplication Columns of **3, 6, 9.**

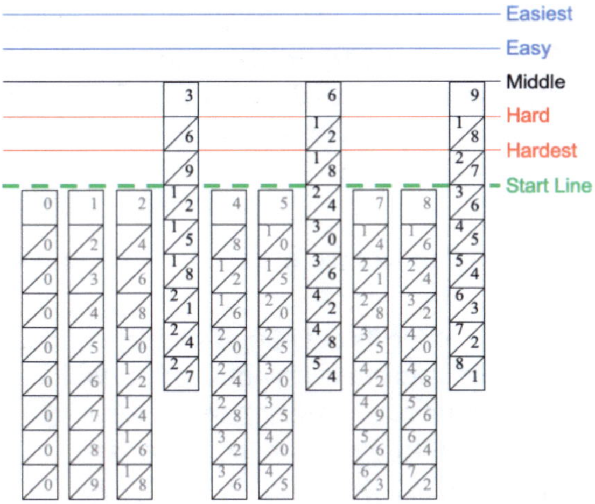

Figure 3.4.1-7: Easy to Hard – 3,6,9.

Three: does not have a rhythm to it that I know of, so we can put it on the Middle line because the unique feature here is the pattern of the products (3, 6, 9).

3s method 1: The digital summation of the 10s and 1s values can always produce the same repeating products.

Sequence:	3, 6, 9,	12, 15, 18,…..	21, 24, 27
digital root:	3, 6, 9,	3, 6, 9,	3, 6, 9

Another feature of the 3s column as noted in the 5s column is that 10 divided by 3 is 3 with a single remainder. So, the repeat sequence of 10s place starts a 0,0,0, then 1,1,1, and 2,2,2.

3s method 2:

3, 6, 9, 12, 15, 18, 21, 24, 27, 30, 33, 36z

Note: this same repeating of the single-digit values is also found in the 6s and 9s columns.

Six: Uses two of the patterns of the three above:

6s method 1: The digital summation of the 10s and 1s values can always produce the same repeating products.

Sequence:	6,	12,	18,	24,	30,	36,…..	42,	48,	54
digital root:	6,	3,	9,	6,	3,	9,	6,	3,	9

6s method 2: The Egyptians noticed, as in the 2 and 4 sequences before, that 6 is the double of 3, and using the 3s stick sequence, doubling put one value between each of the 3s stick sequences.

3, **6**, 9, **12**, 15, **18**, 21, **24**, 27, **30**, 33, **36**, 39, **42**, 45, **48**, 51, **54**

This is the triple of the 3s column of multiplication. As explained earlier, this means that the 3, 6, 9 patterns are taken every second value. The first from the 3,6,9 is 6, then jump two for 3, and just two for 9. So, the pattern is not 3, 6, 9 instead, it is the first three products of the 6s or 6, 3,9 as shown above in the digital root.

Nine: does not have a rhythm like the twos or fives stick; however, it does have a 2-patterns:

9s method 1: Is the triple of the 3s column of multiplication, which means that the 3, 6, and 9 pattern is taken every third value, which in this case is always 9, and therefore each row must sum to the value of 9.

Sequence:	9,	18,	27,	36,	45,	54,	63,	72,	81
digital root:	9,	9,	9,	9,	9,	9,	9,	9,	9

9s method 2: Commonly referred to as the 'Magic 9.'

3, 6, **9**, 12, 15, **18**, 21, 24, **27**, 30, 33, **36**, 39, 42, **45**, 48, 51, **54**, 57, 60, **63**

You will notice that, for tripling of three, there are two values between each multiple value of 9. Look at the digital root of each 9s value: 1) it is a count-down pattern (9 – 1) and 2) it is a repeating digital root of 9. Now, look at the 10s place values from top to bottom; it is a count-up (0 -8). This is why the 9s method 2: Commonly referred to as the 'Magic 9' implies something not mathematical. But, not covered in that 'Magic 9' statement is a new rule that:

For 9 multiplication, the 10s value is always one less than the multiplier.

i.e., 9x2 is 18 where for the 10s value of 2, the 10s value in the answer is 2-1 or 1. Combining that with the sum of the 10s and 1s value must be 9, which means the 1s value is 8. This part of 'The Magic 9' and its relationship with 3 is never shown in school.

3.4.1.7 The Multiplication Column of Seven (7).

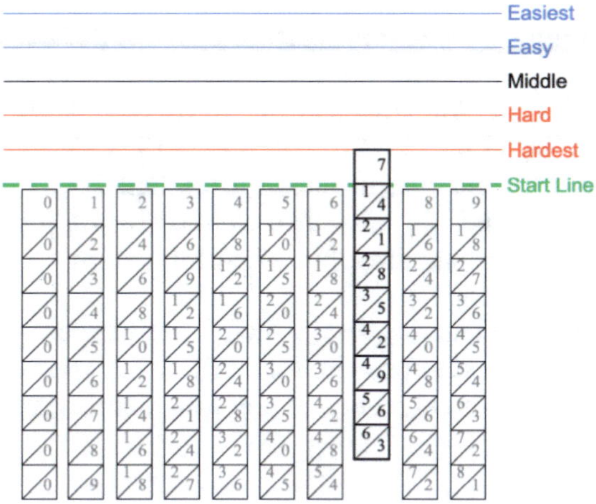

Figure 3.4.1-8: Easy to Hard – Seven (7).

Seven: is the hardest multiplication table because it has no rhythm or simple pattern.

For now, you may have to memorize this one multiplication sequence. But, do notice that 1) the 10s value is a sequence of 0 to 6, except for the double 2 and 4; 2) the digital root is 7, 5, 3, 1, 8, 6, 4, 2, 9; where the 7, 8, 9 are in sequence; as are the 5, 3,1 odd and 6, 4,2 even sequence; and 3) that the pattern is alternating **even** and odd values: **0**, 7, **14**, 21, **28**, 35, **42**, 49, **56**, 63.

In the third example: The even pattern is a cycle starting with 0 and increasing in the 1s value by *4*. The odd pattern is a cycle starting with 7 and also increasing by *4*. So, the only thing you need to memorize is the first two values are 0,7; and that all others are +4 for each 1 and 7 in that order.

1s value: 0, 7, (0+4), (7+4), (0+8), (7+8), (0+12), (7+12), (0+16), (7+16)
10s and 1s values: 0, 7, a+4, b+11, c+8, d+15, e+12, f+19, g+16, h+23
Solution: **0**, 7, a+4=**14**, b+11=21, c+8=**28**, d+15=35, e+12=**42**, f+19=49, g+16=**56**, h+23=63

So, the 10s values are: a=10, b=10, c=20, d=20, e=30, f=30, g=40, h=40. Note that you have a=b=10, c=d=20, e=f=30, and g=h=40. This is a little complex, but the ideal *Rule for 7s* is:

1) Using 0, 7 as the first even and odd numbers, the 1s value increases by 4 for both 0 and 7.
2) 10s values are doubles of 0 for 0 and 7 and increases by 10 for each following pairs.

The first cycle of the. 7s value repeats from 0, 7, 14, 21, 28, 35, 42, 49, 56, 63, as such it will repeat for the 10s value starting with 10 x7 or 70. Which is the next sum of 63 + 7 or 70. Now starting with 7 * 10 = 70, you add 70 to each initial value: 70, 77, 84, 91, 98, 105, 112, 119, 126, 133. You do the same for 7 * 20 = 140, and even sequence after that: 210, 280, 350, 420, 490, 560, and 620. You see the 10s value also went up by 10, just as the 1s value goes up. Again, not understanding, but a helpful pattern.

I know that this is more like Algebra, which I don't expect you to understand. So, in the next Pattern in Section 3.4.2.1, you will see a sum pattern that saves you from algebra and any rule memorization.

3.4.1.8 Summary.

Take your time, and look over the pattern of each stick value. Three patterns are easy: Zero (0), One (1), Five (5), and the next two involve doubles or triples: Two (2, 4, 8), Three (3, 6, 9). This means you don't have to memorize ten multiplication tables and any greater than nine.

Only the seventh needs a new approach (i.e., Pattern). You can see that in Pattern 3 in the next section.

3.4.2: Pattern 2 – Addition of Multiplication Sticks.

Another way to reproduce the pattern of the multiplication table is to use addition, which is the building block of multiplication. This pattern is not taught in school; however, the proof is very straightforward. See reference to the common solution of two addends.

3.4.2.1 Addition of Multiplication Columns.

Any value, or a multiplication column, can be broken down into any two numbers that are equal to or less than the value, which is basic addition. It can also be interpreted as a prelude to subtraction. For example, from the previous "Easy to Hard" pattern, we found we could not find a pattern for the 7s multiplication column. If we break the value 7 into any two numbers that sum to the value 7, we get Figure 3.4.2-1.

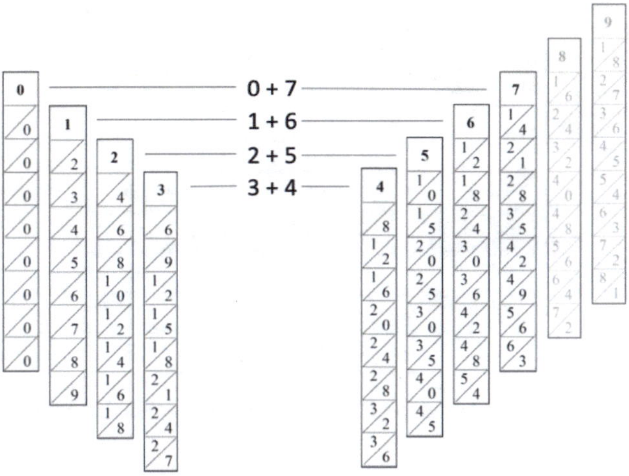

Figure 3.4.2-1: Addition of Multiplication Columns that equals 7.

So, 0 +7, 1 + 6, 2 + 5, and 3 + 4, all add up to 7. So too are their opposites: 7 + 0, 6 + 1, 5+ 2, 4 +3. What is surprising is this is also correct for each row of the associate multiplication sticks. In that, each row adds up to the corresponding row of the 7s multiplication column.

For example, using 2 and 5 Tally-x sticks, see Figure 3.4.2-1, you get Figure 3.4.2-2 below. You will also see that each descending row is equal to the associate 7s multiplication row.

Note: unlike multiplication, the diagonals are not added, but each row value is treated as a whole, i.e., the 8th row yields 16 + 40 = 56.

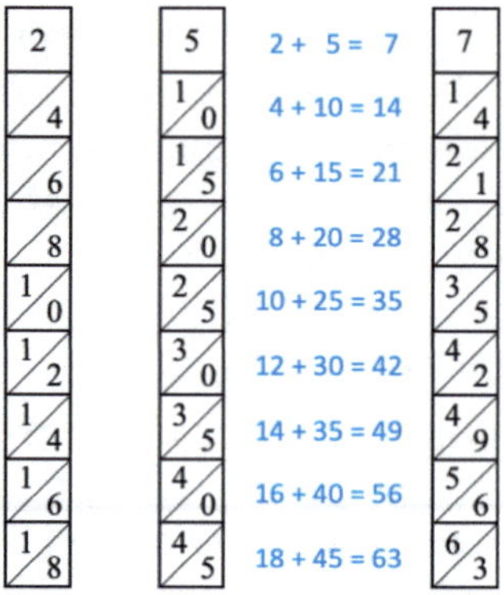

Figure 3.4.2-2 Example of Addition of Multiplication Columns that equals 7 (i.e., 2 + 5).

This can also be done without the sticks, using the 7s table in Figure 3.4.2-3:

2	+	5	=	7
4	+	10	=	14
6	+	15	=	21
8	+	20	=	28
10	+	25	=	35
12	+	30	=	42
14	+	35	=	49
16		40		56
18	+	45	=	63

Figure 3.4.2-3: Addition of Multiplication Columns that equals 7.

3.4.2.2 Other Addition of Columns.

	1	+	1	=	2
0	0	+	0	=	0
1	1	+	1	=	2
2	2	+	2	=	4
3	3	+	3	=	6
4	4	+	4	=	8
5	5	+	5	=	10
6	6	+	6	=	12
7	7	+	7	=	14
8	8	+	8	=	16
9	9	+	9	=	18

	1	+	3	=	4
0	0	+	0	=	0
1	1	+	3	=	4
2	2	+	6	=	8
3	3	+	9	=	12
4	4	+	12	=	16
5	5	+	15	=	20
6	6	+	18	=	24
7	7	+	21	=	28
8	8	+	24	=	32
9	9	+	27	=	36

	2	+	4		6
0	0	+	0		0
1	2	+	4		6
2	4	+	8		12
3	6	+	12		18
4	8	+	16		24
5	10	+	20		30
6	12	+	24		36
7	14	+	28		42
8	16	+	32		48
9	18	+	36		54

	3	+	5	=	8
0	0	+	0	=	0
1	3	+	5	=	8
2	6	+	10	=	16
3	9	+	15	=	24
4	12	+	20	=	32
5	15	+	25	=	40
6	18	+	30	=	48
7	21	+	35	=	56
8	24	+	40	=	64
9	27	+	45	=	72

	1	+	2	=	3
0	0	+	0	=	0
1	1	+	2	=	3
2	2	+	4	=	6
3	3	+	6	=	9
4	4	+	8	=	12
5	5	+	10	=	15
6	6	+	12	=	18
7	7	+	14	=	21
8	8	+	16	=	24
9	9	+	18	=	27

	2	+	3	=	5
0	0	+	0	=	0
1	2	+	3	=	5
2	4	+	6	=	10
3	6	+	9	=	15
4	8	+	12	=	20
5	10	+	15	=	25
6	12	+	18	=	30
7	14	+	21	=	35
8	16	+	24	=	40
9	18	+	27	=	45

	3	+	4	=	7
0	0	+	0	=	0
1	3	+	4	=	7
2	6	+	8	=	14
3	9	+	12	=	21
4	12	+	16	=	28
5	15	+	20	=	35
6	18	+	24	=	42
7	21	+	28	=	49
8	24	+	32	=	56
9	27	+	36	=	63

	4	+	5	=	9
0	0	+	0	=	0
1	4	+	5	=	9
2	8	+	10	=	18
3	12	+	15	=	27
4	16	+	20	=	36
5	20	+	25	=	45
6	24	+	30	=	54
7	28	+	35	=	63
8	32	+	40	=	72
9	36	+	45	=	81

Figure 3.4.2-4: Examples of the Addition of Multiplication Columns.

Figure 3.4.2-4 shows several other examples of Adding Multiplications to get 2, 4, 6, 8, 2, 5, 7, and 9. This can be done for any number, including numbers greater than 9. However, this book will stay with the basic 0 – 9 values.

Another example and way to present it is shown in Figure 3.4.2-5 for the 9s multiplication column.

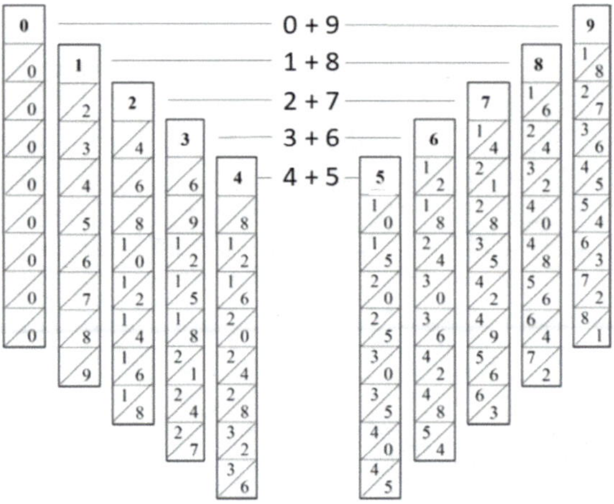

Figure 3.4.2-5: Addition of Multiplication Columns that Equal 9.

For example, using 3 + 6 Tally-x sticks, see Figure 3.4.2-6, we find that each successive row equals the corresponding 9s multiplication Tally-x stick.

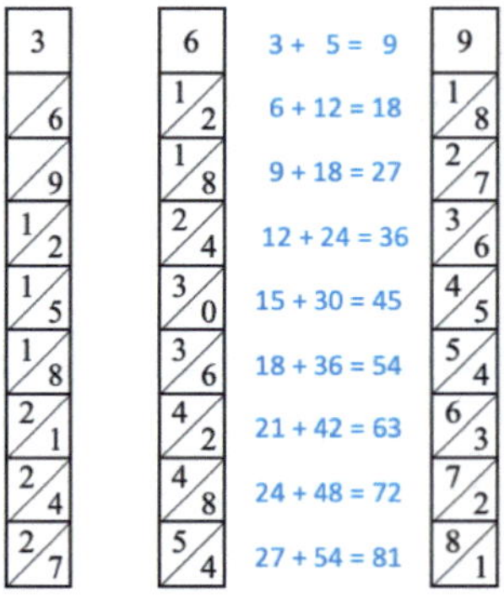

Figure 3.4.2-6: Example of Addition of Multiplication Columns that equals 9.

3.4.2.3 Summary.

So, the first Pattern of Easy to Hard worked fine for all Tally-x sticks (0, 1, 2, 3, 4, 5, 6, 8, and 9), but it did not work for the 7s Tally-x stick. Not only does this Pattern 2 provide for the 7s Tally-x stick, see Figure 3.4.2.1-2, but it can also be used for all the Tally-x sticks above the value of 1.
But, it can also be used to create larger multiplicand values. For example, 5 +5 = 10, 4 + 7 = 11, and 4 + 8 = 12.

In 3.4.3, Pattern 3, you may find an even easier way to create a larger multiplicand.

See *reference 3.5* to use these complimentary values in multiplication, where if the 1s values are complementary numbers, the 1s value is the multiplication of the complementary numbers. Where 10s or leading numbers are the same, then take the leading number without its rightmost digit, and multiply it by the next higher whole number. I.e., 42 x48 is 4 x5 or 20 and the right number is 2 x8 or 16. Combined they form 2016. This is a pattern but also memory learning.

3.4.3: Pattern 3 – Multi-digit Multiplicand Numbers.

All the patterns have to do with single-digit multiplicand values (column), and single-digit multipliers (rows), except for what was discussed in the previous Pattern 2 Summary.

3.4.3.1 Continuing Beyond the 0 through 9 Tally-x sticks.

For Figure 3.4.3-1, the Tally-x sticks for 1 and 0 are laid next to each other to get a Multiplicand (column sticks) for the number 10. It is the row that creates the single-digit Multiplier, and one can easily see that this is the Multiplication set for ten. For example, 10 x1 = 10, 10 x4 = 40, and 10 x7 = 70. Where the diagonal values are of the same place value and therefore summed. However, in this case, all values are in the lower right corner of the tile, and therefore, there are no carryovers to be concerned with. This is very easy to visualize and helps in the next value.

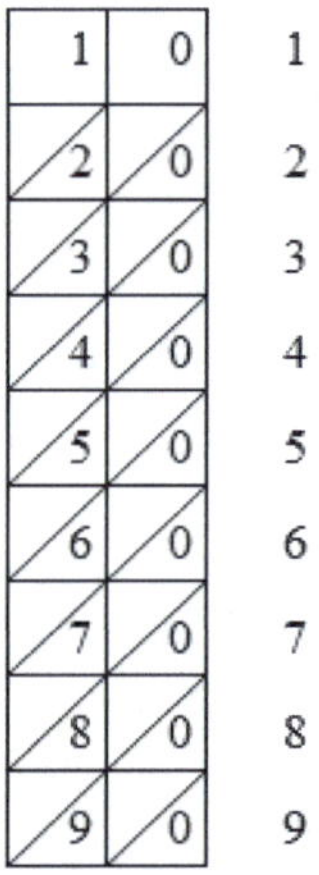

Figure 3.4.3-1: Ten (10) Multiplicand.

The next value is eleven, so Figure 3.4.3-2 is created using a duplicate of the 1s Tally-x sticks laid next to each other to get Multiplicand (column sticks) for the number 11. It is the row that creates the single-digit Multiplier. For example, 11 x1 = 11, 11 x4 = 44, and 11 x7 = 77, where each multiplier is the row count. In this case, the row count is also the 10s stick. Again, this is easily seen and makes this Pattern very understandable. However, what of the number 12?

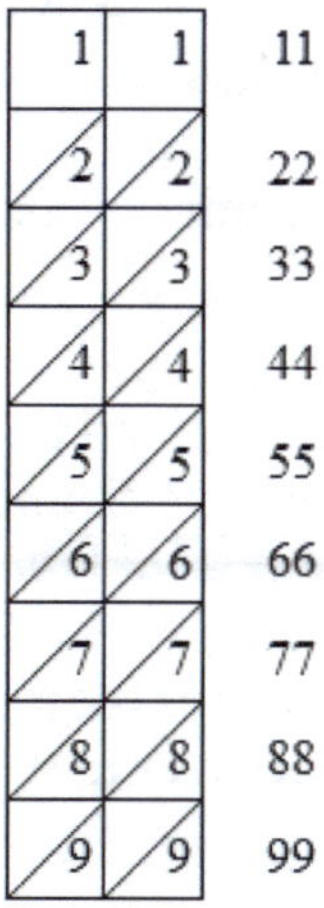

Figure 3.4.3-2: Eleven (11) Multiplicand.

In Figure 3.4.3-3 uses the Tally-x sticks for 1 and 2, which are laid next to each other to get a Multiplicand for the number 12. However, in this case, there can be carryovers from one place value to the next left Tally-x stick. For example, 12 x1 = 12, 12 x4 = 48, and 12 x7 = (10 x7) + (2 x7) = 70 + 14 = 84. Where the diagonal values that are of the same place value 7 and 1 are summed, as they are both of the 10s place value. All carryovers are passed over to the left to the next place value to be summed.

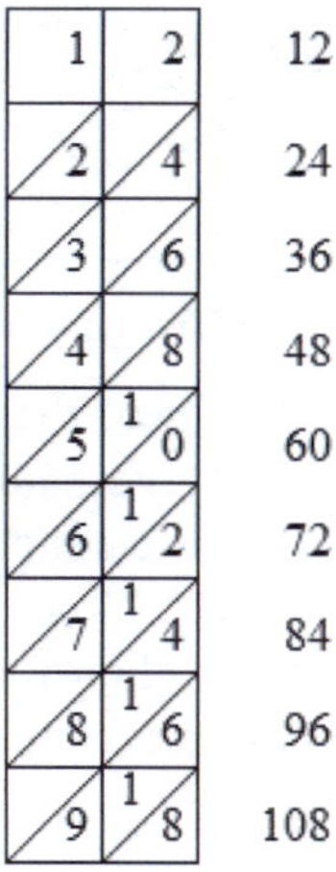

Figure 3.4.3-3: Twelve (12) Multiplicand.

Only 1 2x 9 has a 100s place value carry over because the 10s place value is 9 + 1 or 10; therefore, there are ten 10s values. In a better way to look at it, there is 8 in the 1s place value, 0 (zero) in the 10s place value, and the carryover of 1 to the 100s place value.

3.4.3.2 Three (3) or More Digit Multiplicands.

Just as above, where we can add a Tally-x stick next to the single-digit multiplicand stick to get a 2-digit multiplicand number, so you can continue with a third stick to a 2-place multiplicand to get a 3-place multiplicand, see Figure 3.4.3-4. The Tally-x sticks for 2, 3, and 6 are laid next to each other to get the Multiplicand for the number 236. It is the row that creates the single-digit Multiplier. For example, 236 x1 = 236, 236 x4 = 944, and 236 x7 = 1652. Where the diagonal values are of the same place value and therefore summed. All carryovers are passed over to the left to the next place value to be summed.

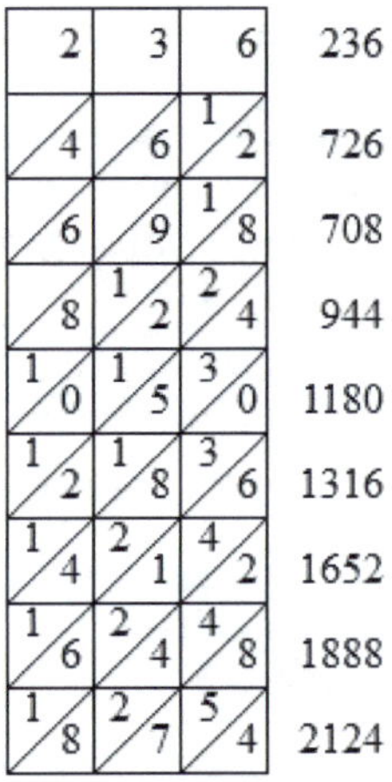

Figure 3.4.3-4: Three (3) digit Multiplicand – 236.

The same is true for Figure 3.4.3-5, where the four Tally-x sticks of 5, 2, 3, and 6 are laid next to each other to get the Multiplicand value of 5236., and again, it is the row that creates the single-digit Multiplier. For example, 5236x1 = 5236, 5236x4 = 20944, and 5236x7 = 36652. Where the diagonal values are of the same place value and therefore summed, and all carryovers are passed over to the left to the next place value to be summed.

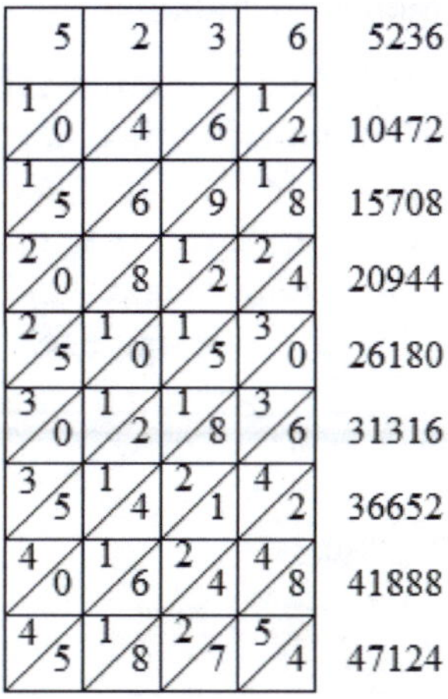

Figure 3.4.3-5: Four (4) digit Multiplicand – 5236.

3.4.3.3 Summary.

As it can be seen, every possible multi-digit Multiplicand number can be created, given enough duplicate Tally-x sticks. So, the limitation of memorizing 0 to 12 can be replaced with Pattern 3 into a never-ending multi-digit Multiplicand number, up in the billions or larger.

This is the beauty of the Tally-x sticks, and this adventure is only beginning.

3.4.4: Pattern 4 – Complementary Columns.

Similar to Pattern 3: Addition of Multiplications, this pattern does not add up to the product multiplication column. In this case, the complementary multiplication columns add up to ten (10). In this case, that would be 9 and 1, 8 and 2, 7 and 3, and 6 and 4. As such, the 1s values are inverts of each other. The 5 and 5 do not work as they are duplicates and are the inverted 1s place value of each other. You will also notice that each row sum of the 9s multiplication and 1s multiplication adds up in each sequential 10s value: row 1 is 10, row 2 is 20, row 3 is 30, …, row 9 is 90.

3.4.4.1 Complementary of 9s and 1s.

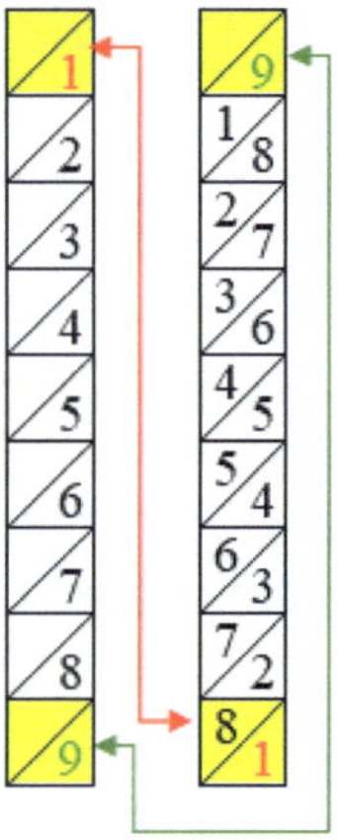

Figure 3.4.4-1: Complementary Columns – 9 and 1.

Table 3.4.4-1 shows, based on the "Magic 9" feature, that in the 9s multiplication column, you have a count-down from 9 to 1 in the 1s place value. Conversely, the 1s multiplication column is a count-down from 1 to 9 in the 1s place value.

1s & 9s

1						1s	9						1s
	1	x	1	=	1	1		9	x	1	=	9	9
	1	x	2	=	2	2		9	x	2	=	18	8
	1	x	3	=	3	3		9	x	3	=	27	7
	1	x	4	=	4	4		9	x	4	=	36	6
	1	x	5	=	5	5		9	x	5	=	45	5
	1	x	6	=	6	6		9	x	6	=	54	4
	1	x	7	=	7	7		9	x	7	=	63	3
	1	x	8	=	8	8		9	x	8	=	72	2
	1	x	9	=	9	9		9	x	9	=	81	1

Table 3.4.4-1: Examples of Complementary Columns – 9 and 1.

3.4.4.2 Complementary of 2s and 8s.

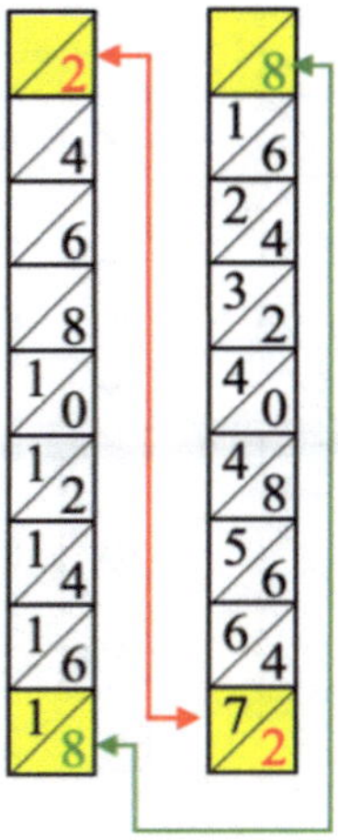

Figure 3.4.4-2: Complementary Columns – 2 and 8.

Table 3.4.4-2 shows, based on the "Even" numbers, that in the 8s multiplication column, you have a count-down from 2 to 8 twice in the 1s place value. Conversely, the 2s multiplication column is a count-down from 2 to 8 in the 1s place value.

2s & 8s

2					1s	8					1s
2	x	1	=	2	2	8	x	1	=	8	8
2	x	2	=	4	4	8	x	2	=	16	6
2	x	3	=	6	6	8	x	3	=	24	4
2	x	4	=	8	8	8	x	4	=	32	2
2	x	5	=	10	0	8	x	5	=	40	0
2	x	6	=	12	2	8	x	6	=	48	8
2	x	7	=	14	4	8	x	7	=	56	6
2	x	8	=	16	6	8	x	8	=	64	4
2	x	9	=	18	8	8	x	9	=	72	2

Table 3.4.4-2: Examples of Complementary Columns –2 and 8.

3.4.4.3 Complementary of 3s and 7s.

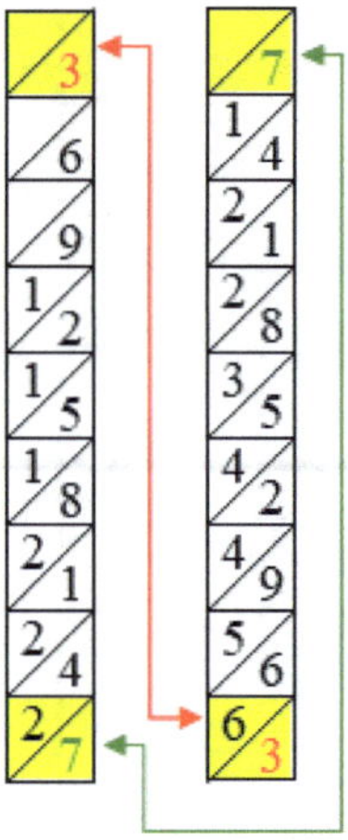

Figure 3.4.4-3: Complementary Columns – 3 and 7.

Table 3.4.4-3 shows, based on the "odd" numbers, that in the 7s multiplication column, you have a count-down from 7 in the 1s place value, but you skip every 7th value. Conversely, the 3s multiplication column is a count-down from 3 in the 1s place value, but you skip every 3rd value.

3s & 7s

3						1s	**7**						1s
	3	x	1	=	3	3		7	x	1	=	7	7
	3	x	2	=	6	6		7	x	2	=	14	4
	3	x	3	=	9	9		7	x	3	=	21	1
	3	x	4	=	12	2		7	x	4	=	28	8
	3	x	5	=	15	5		7	x	5	=	35	5
	3	x	6	=	18	8		7	x	6	=	42	2
	3	x	7	=	21	1		7	x	7	=	49	9
	3	x	8	=	24	4		7	x	8	=	56	6
	3	x	9	=	27	7		7	x	9	=	63	3

Table 3.4.4-3: Examples of Complementary Columns – 3 and 7.

3.4.4.4 Complementary of 4s and 6s.

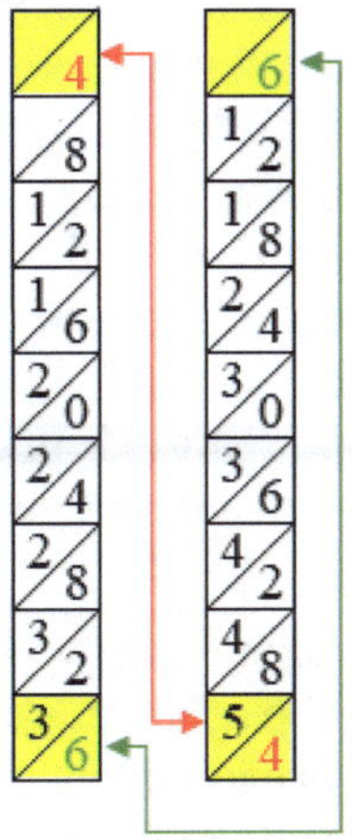

Figure 3.4.4-4: Complementary Columns – 4 and 6.

Table 3.4.4-4 shows, based on the "even" numbers, that in the 6s multiplication column, you have a count-down from 6 in the 1s place value, but you skip every 6th value. Conversely, the 4s multiplication column is a count-down from 4 in the 1s place value; however, you skip every 4th value.

4s &6s

4					1s	6					1s
4	x	1	=	4	4	6	x	1	=	6	6
4	x	2	=	8	8	6	x	2	=	12	2
4	x	3	=	12	2	6	x	3	=	18	8
4	x	4	=	16	6	6	x	4	=	24	4
4	x	5	=	20	0	6	x	5	=	30	0
4	x	6	=	24	4	6	x	6	=	36	6
4	x	7	=	28	8	6	x	7	=	42	2
4	x	8	=	32	2	6	x	8	=	48	8
4	x	9	=	36	6	6	x	9	=	54	4

Table 3.4.4-4: Examples of Complementary Columns – 4 and 6.

3.4.4.5 Complementary of 5s.

Multiplication Column of Five (5) states that the number of repeating sequences can be found in the 1s place value. In the 5s multiplication column, you can have a count-down of 0 and 5 that repeats 5x in the 1s place value. However, the 10s place will repeat twice: 0, 0; 1, 1; 2, 2; … 9, 9.

With regards to the 1s place, you can find that it is "not" complementary to itself, as stated in the opening of this chapter.

3.4.4.6 Summary.

This pattern, which isn't typically taught in schools or colleges, becomes clear once it's revealed. It suggests that memorizing patterns from 0 to 9 isn't essential; focusing on 0 to 5 is sufficient, as 6 to 9 simply reflect the 1s place values of 4 to 1. As mentioned in Section 3.4.1.1, the multiplication table for zero (0) is too basic, while Section 3.4.1.4 discusses the straightforward nature of the five (5) table. Here, our focus is on 1 to 4, and their 1s place complements of 6 to 9. Reviewing tables 3.4.4-1 to 3.4.4-5 reveals this simple pattern, with the challenge mainly lying in the 10s place values. It's truly remarkable how mathematics is based on patterns found in nature as well.

It's worth noting that while complementarity is horizontal, each single digit is a 1s place value complement to the other; for double digits, the 1s place values are also complementary to each other. Similarly, like the single-digit values mentioned earlier, each row represents sequential hundreds: row 1 is 100, row 2 is 200, row 3 is 300, and so on up to row 9 being 900. You can try extending this pattern yourself to 100 and even 1000.

What a marvel it is that mathematics is built upon patterns that can also be observed in nature!

3.4.5: Pattern 5 –Across a Stick.

It should be noted that there is a similarity between Pattern 5 and Pattern 4: Complementary Columns. It could have just as easily been called Complementary stick rows. The reason is that just as Pattern 4: Complementary Columns has a counterpart stick horizontally, i.e., 9, and 1, so Pattern 5 has a counterpart within each multiplicand stick, top to bottom.

3.4.5.1 Introduction.

In Figure 3.4.5-1, we see that the difference is the multiplier five from value-1 on the top, to value-2 on the bottom, 1 x5 = 5.

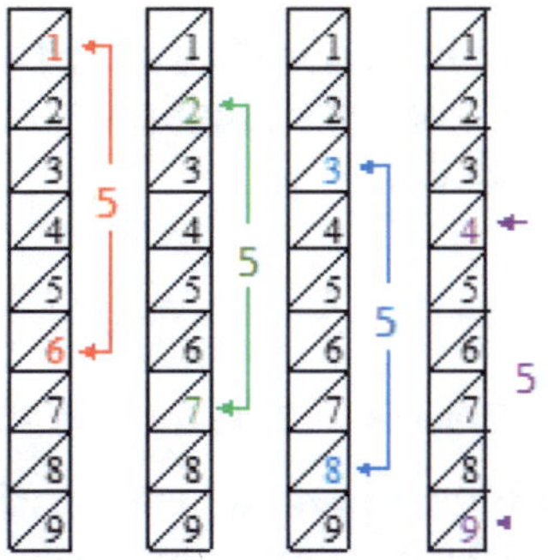

Figure 3.4.5-1: Across a Stick – 1-multiplier.

In Table 3.4.5-1, we have the multiplication table for the 1s multiplicand (column values) for the first five rows (multipliers) in red on the top row and the next five rows underneath each.

1

1 x 1 = 1	1 x 2 = 2	1 x 3 = 3	1 x 4 = 4	1 x 5 = 5
1 x 6 = 6	1 x 7 = 7	1 x 8 = 8	1 x 9 = 9	1 x 10 = 10

Table 3.4.5-1: Examples of Across a Stick – 1-multiplier.

Note the difference between the lower and upper red values is 5, and the difference in the answer is 5 (or 1 x5). In Figure 3.4.5-2, the difference is the multiplier time five, 2 x5 = 10.

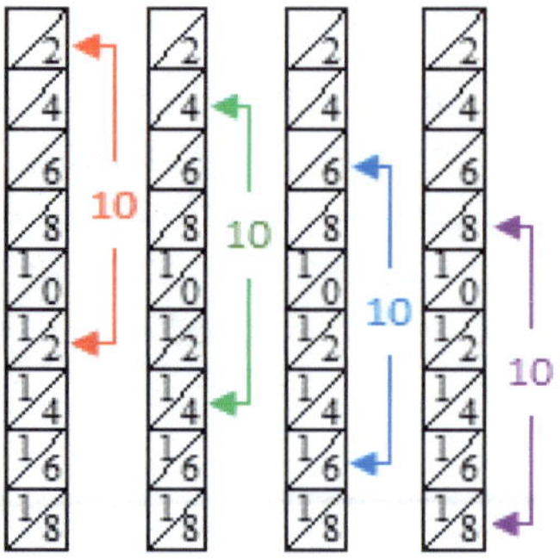

Figure 3.4.5-2: Across a Stick – 2-multiplier.

In Table 3.4.5-2, we have the multiplication table for the 2s multiplicand (column values) for the first five rows (multipliers) in red on the top row and the next five rows underneath each.

2

2 x 1 = 2	2 x 2 = 4	2 x 3 = 6	2 x 4 = 8	2 x 5 = 10
2 x 6 = 12	2 x 7 = 14	2 x 8 = 16	2 x 9 = 18	2 x 10 = 20

Table 3.4.5-2: Examples of Across a Stick – 2-multiplier.

Note the difference between the lower and upper red values is 5, and the difference in the answer is 10 (or 2 x5). In Figure 3.4.5-3, the difference is the multiplier time five, 3 x5 = 15.

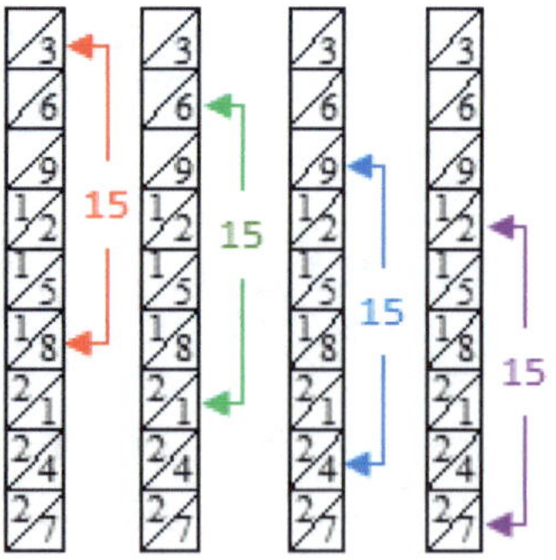

Figure 3.4.5-3: Across a Stick – 3-multiplier.

In Table 3.4.5-3, we have the multiplication table for the 3s multiplicand (column values) for the first five rows (multipliers) in red on the top row and the next five rows underneath each.

3

3 x 1 = 3	3 x 2 = 6	3 x 3 = 9	3 x 4 = 12	3 x 5 = 15
3 x 6 = 18	3 x 7 = 21	3 x 8 = 24	3 x 9 = 27	3 x 10 = 30

Table 3.4.5-3: Examples of Across a Stick – 3-multiplier.

Note the difference between the lower and upper red values is 5, and the difference in the answer is 15 (or 3 x5). In Figure 3.4.5-4, the difference is the multiplier time five, 4 x5 = 20.

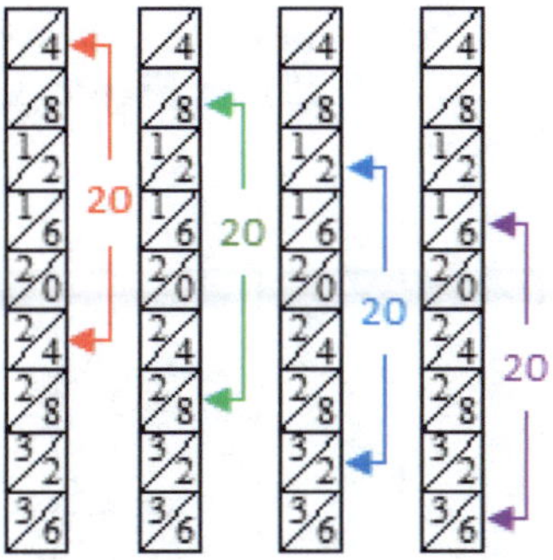

Figure 3.4.5.4: Across a Stick – 4-multiplier.

In Table 3.4.5-4, we have the multiplication table for the 4s multiplicand (column values) for the first five rows (multipliers) in red on the top row and the next five rows underneath each.

4

4	x	1	=	4	4	x	2	=	8	4	x	3	=	12	4	x	4	=	16	4	x	5	=	20
4	x	6	=	24	4	x	7	=	28	4	x	8	=	32	4	x	9	=	36	4	x	10	=	40

Table 3.4.5-4: Examples of Across a Stick – 4-multiplier.

Note the difference between the lower and upper red values is 5, and the difference in the answer is 20 (or 4 x5). In Figure 3.4.5-5, the difference is the multiplier time five, 5 x5 = 25.

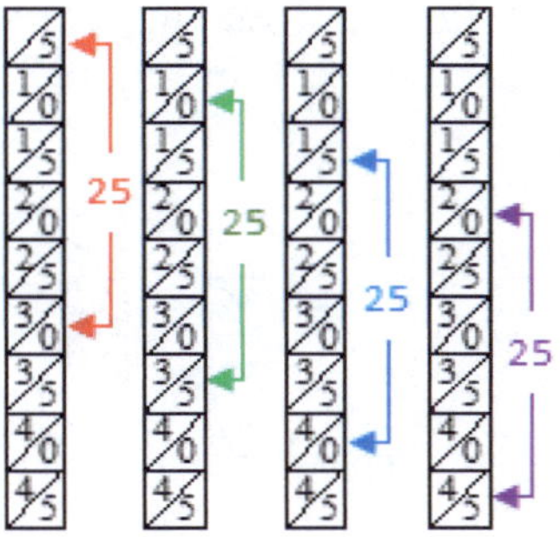

Figure 3.4.5-5: Across a Stick – 5-multiplier.

In Table 3.4.5-5, we have the multiplication table for the 5s multiplicand (column values) for the first five rows (multipliers) in red on the top row and the next five rows underneath each.

5

5	x	1	=	5	5	x	2	=	10	5	x	3	=	15	5	x	4	=	20	5	x	5	=	25
5	x	6	=	30	5	x	7	=	35	5	x	8	=	40	5	x	9	=	45	5	x	10	=	50

Table 3.4.5-5: Examples of Across a Stick – 5-multiplier.

Note the difference between the lower and upper red values is 5, and the difference in the answer is 25 (or 5 x5). In Figure 3.4.5-6, the difference is the multiplier time five, 6 x5 = 30.

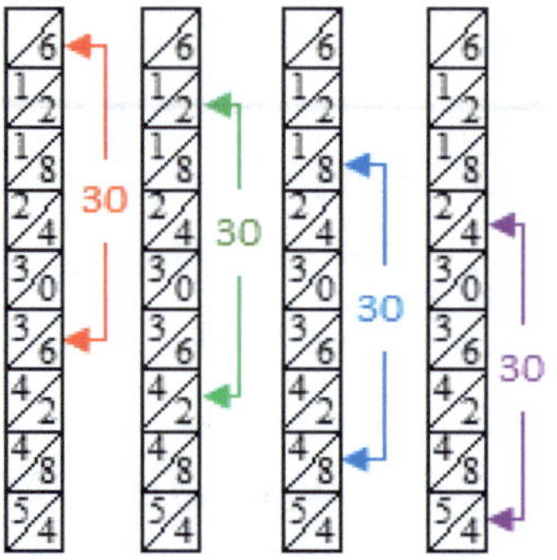

Figure 3.4.5-6: Across a Stick – 6-multiplier.

In Table 3.4.5-6, we have the multiplication table for the 6s multiplicand (column values) for the first five rows (multipliers) in red on the top row and the next five rows underneath each.

6

6	x	1	=	6	6	x	2	=	12	6	x	3	=	18	6	x	4	=	24	6	x	5	=	30
6	x	6	=	36	6	x	7	=	42	6	x	8	=	48	6	x	9	=	54	6	x	10	=	60

Table 3.4.5-6: Examples of Across a Stick – 6-multiplier.

Note the difference between the lower and upper red values is 5, and the difference in the answer is 30 (or 6 x5). In Figure 3.4.5-7, the difference is the multiplier time five, 7 x5 = 35.

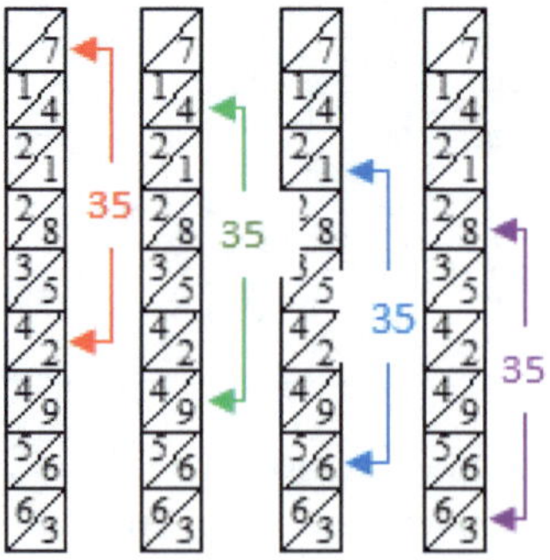

Figure 3.4.5-7: Across a Stick – 7-multiplier.

In Table 3.4.5-7, we have the multiplication table for the 7s multiplicand (column values) for the first five rows (multipliers) in red on the top row and the next five rows underneath each.

7

7 x 1 = 7	7 x 2 = 14	7 x 3 = 21	7 x 4 = 28	7 x 5 = 35
7 x 6 = 42	7 x 7 = 49	7 x 8 = 56	7 x 9 = 63	7 x 10 = 70

Table 3.4.5-7: Examples of Across a Stick – 7-multiplier.

Note the difference between the lower and upper red values is 5, and the difference in the answer is 35 (or 7 x5). In Figure 3.4.5-8, the difference is the multiplier time five, 8 x5 = 40.

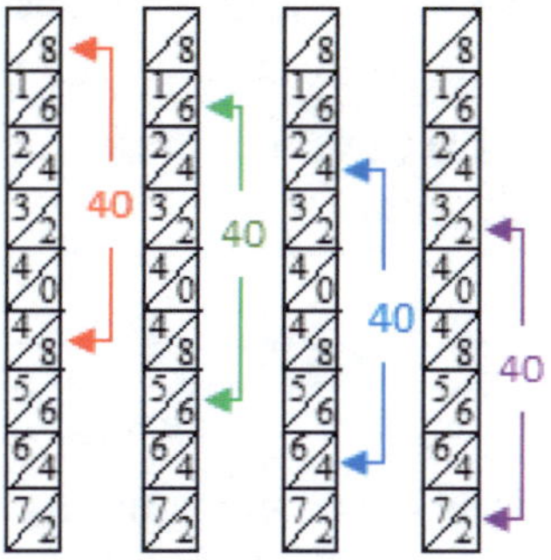

Figure 3.4.5-8: Across a Stick – 8-multiplier.

In Table 3.4.5-8, we have the multiplication table for the 8s multiplicand (column values) for the first five rows (multipliers) in red on the top row and the next five rows underneath each.

8

8 x 1 = 8	8 x 2 = 16	8 x 3 = 24	8 x 4 = 32	8 x 5 = 40
8 x 6 = 48	8 x 7 = 56	8 x 8 = 64	8 x 9 = 72	8 x 10 = 80

Table 3.4.5-8: Examples of Across a Stick – 8-multiplier.

Note the difference between the lower and upper red values is 5, and the difference in the answer is 40 (or 8 x5). In Figure 3.4.5-9, we see that the difference is the multiplier time five, 9 x5 = 45.

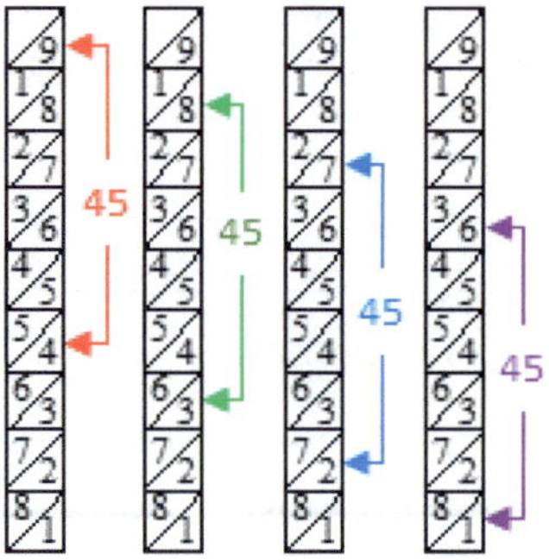

Figure 3.4.5-9: Across a Stick – 9-multiplier.

In Table 3.4.5-9, we have the multiplication table for the 9s multiplicand (column values) for the first five rows (multipliers) in red on the top row and the next five rows underneath each.

9																									
	9	x	1	=	9	9	x	2	=	18	9	x	3	=	27	9	x	4	=	36	9	x	5	=	45
	9	x	6	=	54	9	x	7	=	63	9	x	8	=	72	9	x	9	=	81	9	x	10	=	90

Table 3.4.5-9: Examples of Across a Stick – 9-multiplier.

Note the difference between the lower and upper red values is 5, and the difference in the answer is 45 (or 9 x5). In Figure 3.4.5-10, we see that the difference is the multiplier time five, 10 x5 = 50.

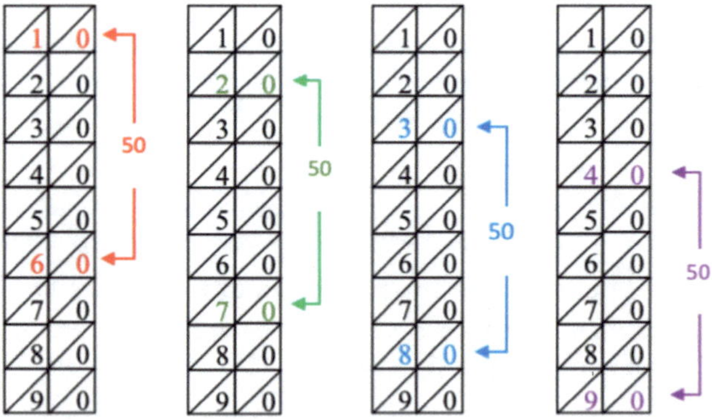

Figure 3.4.5-10: Across a Stick – 10-multiplier.

In Table 3.4.5-10, we have the multiplication table for the 10s multiplicand (column values) for the first five rows (multipliers) in red on the top row and the next five rows underneath each.

10

10 x 1 = 10	10 x 2 = 20	10 x 3 = 30	10 x 4 = 40	10 x 5 = 50
10 x 6 = 60	10 x 7 = 70	10 x 8 = 80	10 x 9 = 90	10 x 10 = 100

Table 3.4.5-10: Examples of Across a Stick – 10-multiplier.

Note the difference between the lower and upper red values is 5, and the difference in the answer is 50 (or 10 x5).

3.4.5.2 Across a Stick that is Greater than the 10-multiplier.

This action will repeat for larger column values (multiplicands); however, we have to remember the columnar stick can be used to create larger numbers above the 1s place value just as done in making numbers greater than 9. For example, 17 is just a 1 in the 10s place value; and a 7 in the 1s place value; or 253 is a 2 in the 100s place value 5 in the 10s place value; and 3 in the 1s place value.

In Figure 3.4.5-11, we see that the difference is the multiplier time five, 17 x5 = 85.

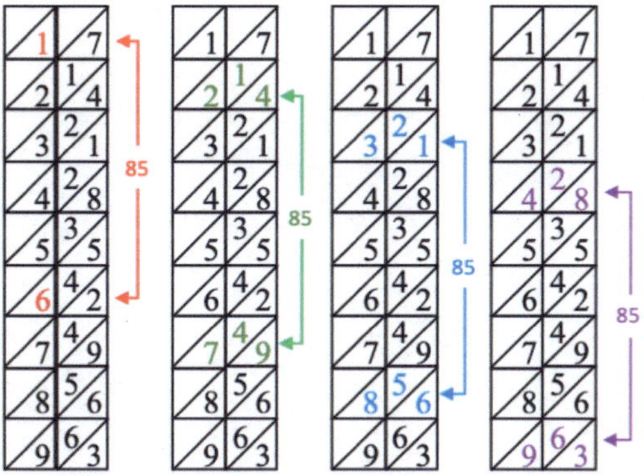

Figure 3.4.5-11: Across a Stick – 17-multiplier.

In Table 3.4.5-11, the number 17 is placed by stacking the 1 in the 10s place value as 1x10, and the 7 placed in the 1s place value as 7x1. Then they are stacked over each other for the first 5 rows (multiplier), and repeated for the next 5 rows.

However, the answer is in the summation of the two (10, 7) shown under the summation double line.

17

17 x 1 = 17	17 x 2 = 34	17 x 3 = 51	17 x 4 = 68	17 x 5 = 85
17 x 6 = 102	17 x 7 = 119	17 x 8 = 136	17 x 9 = 153	17 x 10 = 170

Table 3.4.5-11: Examples of Across a Stick – 17-multiplier.

Note the difference between the Lower and upper blue values is still 5, and the difference in the answers is 85 (or 17 x5).

3.4.5.3 Summary.

This pattern applies universally, regardless of the size of the number. It's an inherent feature of the multiplication table, visible when examining any column up or down. Another notable observation is the consistent five-unit difference. In cases where the multiplicand is odd, the unit place will always be 5. This observation finds support in Pattern 7: the Checkerboard Pattern.

3.4.6 Pattern 6: – Columns and Rows.

This section begins the first of the five-sections associated with the Table based group, where we may learn that patterns are not just vertical or horizontal, but diagonally with complementary rows and columns.

3.4.6.1 Column and Row.

The column and row duplication continues throughout each multiplicand (column value) and multiplier (row value). This is also why the multiplicand and the multiplier can be swapped, and it is called the Commutative Properties of multiplication, i.e., a x b = b x a. In Pattern 6, we are only concerned with multiplicand and multiplier values that are equal.

0-Column and 0-Row:

Figure 3.4.6-1: Column and Row of 0.

In Figure 3.4.6-1, note that the 0s value column and the 0s value row are highlighted in the figure and have the values of the same products in a top-to-bottom, and left-to-right sequence. The red value is the square of the column or row value at the intersection of the common multiplier, i.e., 0 x0 = 0.

Note that whether a column or row value, as you progress horizontally or vertically away from the squared value, each sequence is the same.

1-Column and 1-Row:

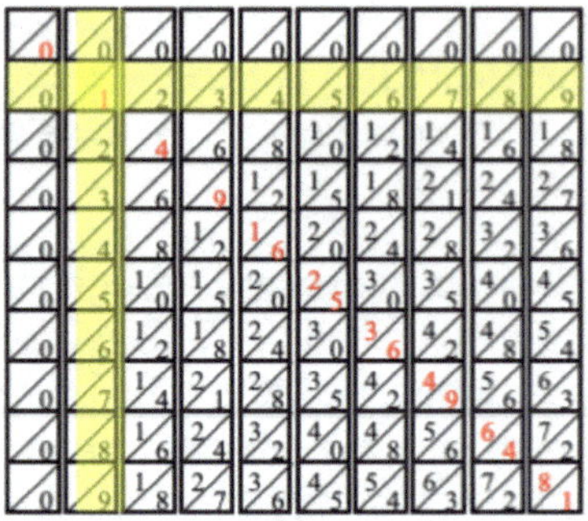

Figure 3.4.6-2: Column and Row of 1.

In Figure 3.4.6-2, note that the 1s value column and the 1s value row are highlighted in the figure and have the values of the same products in a top-to-bottom, and left-to-right sequence. This way you can see that both are the same. The red value is the square of the column or row value at the intersection of the common multiplier, i.e., 1 x1 = 1.

2-Column and 2-Row:

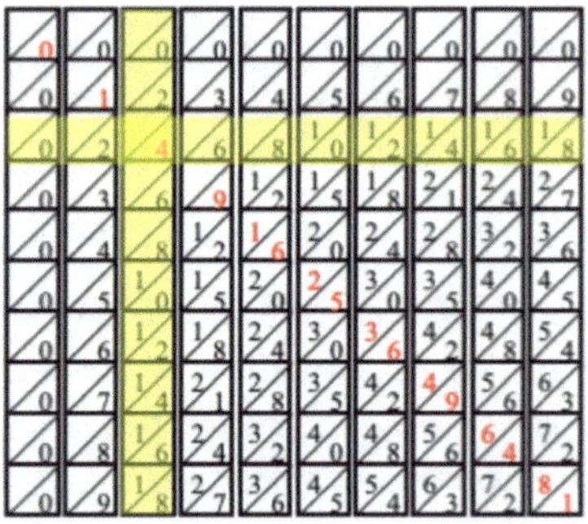

Figure 3.4.6-3: Column and Row of 2.

In Figure 3.4.6-3, note that the 2s value column and the 2s value row are highlighted in the figure and have the values of the same products in a top-to-bottom, and left-to-right sequence. The red value is the square of the column or row value at the intersection of the common multiplier, i.e., 2 x2 = 4.

3-Column and 3-Row:

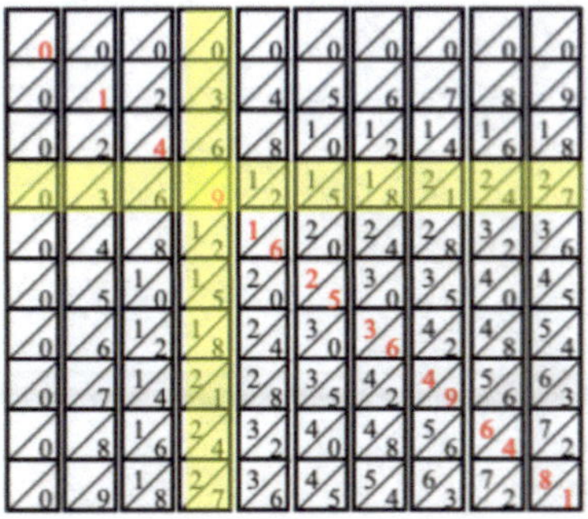

Figure 3.4.6-4: Column and Row of 3.

In Figure 3.4.6-4, note that the 3s value column and the 3s value row are highlighted in the figure and have the values of the same products in a top-to-bottom, and left-to-right sequence. The red value is the square of the column or row value at the intersection of the common multiplier, i.e., 3 x3 = 9.

4-Column and 4-Row:

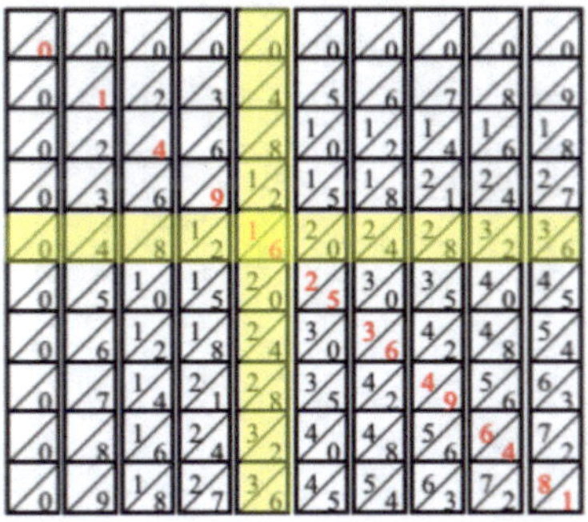

Figure 3.4.6-5: Column and Row of 4.

In Figure 3.4.6-5, note that the 4s value column and the 4s value row are highlighted in the figure and have the values of the same products in a top-to-bottom, and left-to-right sequence. The red value is the square of the column or row value at the intersection of the common multiplier, i.e., 4 x4 = 16.

5-Column and 5-Row:

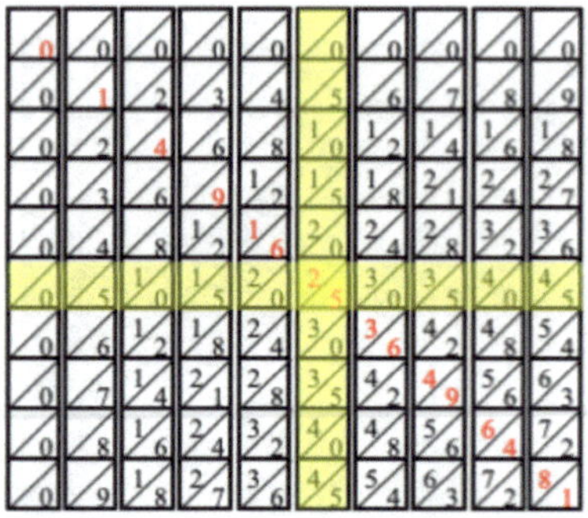

Figure 3.4.6-6: Column and Row of 5.

In Figure 3.4.6-6, note that the 5s value column and the 5s value row are highlighted in the figure and have the values of the same products in a top-to-bottom, and left-to-right sequence. The red value is the square of the column or row value at the intersection of the common multiplier, i.e., 5 x5 = 25.

6-Column and 6-Row:

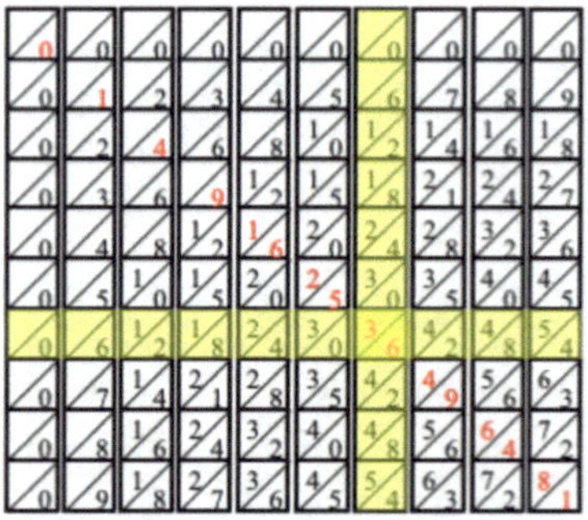

Figure 3.4.6-7: Column and Row of 6.

In Figure 3.4.6-7, note that the 6s value column and the 6s value row are highlighted in the figure and have the values of the same products in a top-to-bottom, and left-to-right sequence. The red value is the square of the column or row value at the intersection of the common multiplier, i.e., 6 x6 = 36. Note that whether a column or row value, as you progress horizontally or vertically away from the squared value, each sequence is the same.

7-Column and 7-Row:

Figure 3.4.6-8: Column and Row of 7.

In Figure 3.4.6-8, note that the 7s value column and the 7s value row are highlighted in the figure and have the values of the same products in a top-to-bottom, and left-to-right sequence. The red value is the square of the column or row value at the intersection of the common multiplier, i.e., 7 x7 = 49.

8-Column and 8-Row:

Figure 3.4.6-9: Column and Row of 8.

In Figure 3.4.6-9, note that the 8s value column and the 8s value row are highlighted in the figure and have the values of the same products in a top-to-bottom, and left-to-right sequence. The red value is the square of the column or row value at the intersection of the common multiplier, i.e., 8 x8 = 64.

9-Column and 9-Row:

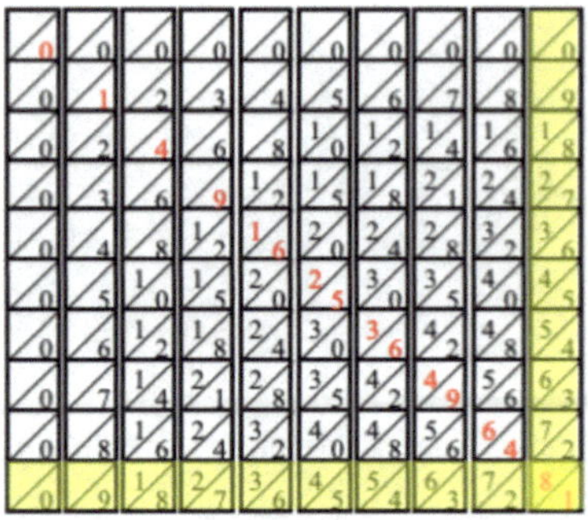

Figure 3.4.6-10: Column and Row of 9.

In Figure 3.4.6-10, note that the 9s value column and the 9s value row are highlighted in the figure and have the values of the same products in a top-to-bottom, and left-to-right sequence. The red value is the square of the column or row value at the intersection of the common multiplier, i.e., 9 x9 = 81.

3.4.6 Summary

This pattern is typically seen by students in grade school where they are required to repeatedly create the multiplication table without mistakes. So, starting with the 0s on the left and on top, then the 9s on the right, and bottom, you circle into the center: 1s then 8s, 2s then 7s …, down to the 4s, and 5s, the last four tiles. As a pattern this results in few or no mistakes but without the understanding behind it, it is forgotten over the years.

3.4.7: Pattern 7 –Checkerboard.

Let us start with even and odd numbers in the Multiplication table, Figure 3.4.7-1, where column (or multiplicand), and row (or multiplier) numbers are at the top, and left side, and the answers (product) are found at the intersection of the column and row numbers. The numbers of the column and row are either even or odd.

A number that can be divided by 2, without a remainder, is an even number: 0, 2, 4, 6, 8. This was discussed in Section 3.4.1.5. In this case, the zero (0) is considered even because it is found halfway between any positive or negative value of the same integer: -2, 0 +2. Any integer that cannot be divided exactly by 2 is an odd number: 1, 3, 5, 7, or 9. You need only look at the 1s place value to determine if a value is even or odd, whether a multiplicand, multiplier, or product.

	1	2	3	4	5	6	7	8	9
1	1	2	3	4	5	6	7	8	9
2	2	4	6	8	10	12	14	16	18
3	3	6	9	12	15	18	21	24	27
4	4	8	12	16	20	24	28	32	36
5	5	10	15	20	25	30	35	40	45
6	6	12	18	24	30	36	42	48	54
7	7	14	21	28	35	42	49	56	63
8	8	16	24	32	40	48	56	64	72
9	9	18	27	36	45	54	63	72	81

Figure 3.4.7-1: Multiplication Table.

When you multiply even or odd numbers the results are either even or odd. However, there is a geometric pattern that can be shown that addresses when:

Even times Even	= Even.
Even times Odd	= Even.
Odd times Even	= Even.
Odd times Odd	= Odd.

So, how do we show this geometrically? We start with statement #1 above, where each even column and even row has an even intersection answer. So, highlight the even Column values and the even Row values. The highlight only shows the intersection values. So, statement #1 says that the multiplication of two even values (i.e., column value and row value) is an even answer, and we can see that in Figure 3.4.7-2.

	1	2	3	4	5	6	7	8	9
1	1	2	3	4	5	6	7	8	9
2	2	4	6	8	10	12	14	16	18
3	3	6	9	12	15	18	21	24	27
4	4	8	12	16	20	24	28	32	36
5	5	10	15	20	25	30	35	40	45
6	6	12	18	24	30	36	42	48	54
7	7	14	21	28	35	42	49	56	63
8	8	16	24	32	40	48	56	64	72
9	9	18	27	36	45	54	63	72	81

Figure 3.4.7-2: Highlight where Even Columns Intersect Even Rows.

Now for statement #2, Even times Odd is Even, let's start with highlighting each even column value, see Figure 3.4.7-3. Looking at the product of any even column value (multiplicand) with any (Even or Odd) row value, the answer is always even. Again, you need only look at the 1s place value to determine if a value is even or odd, whether a multiplicand, multiplier, or product.

	1	2	3	4	5	6	7	8	9
1	1	2	3	4	5	6	7	8	9
2	2	4	6	8	10	12	14	16	18
3	3	6	9	12	15	18	21	24	27
4	4	8	12	16	20	24	28	32	36
5	5	10	15	20	25	30	35	40	45
6	6	12	18	24	30	36	42	48	54
7	7	14	21	28	35	42	49	56	63
8	8	16	24	32	40	48	56	64	72
9	9	18	27	36	45	54	63	72	81

Figure 3.4.7-3: Highlight Even Columns.

If this is true, then by the Commutative Properties for Multiplication, for any two numbers the answer is always the same, no matter the order of the numbers. I.e., 2x3 = 3x2 = 6. As such, this same idea would be true for all even rows, Figure 3.4.7-4.

	1	2	3	4	5	6	7	8	9
1	1	2	3	4	5	6	7	8	9
2	2	4	6	8	10	12	14	16	18
3	3	6	9	12	15	18	21	24	27
4	4	8	12	16	20	24	28	32	36
5	5	10	15	20	25	30	35	40	45
6	6	12	18	24	30	36	42	48	54
7	7	14	21	28	35	42	49	56	63
8	8	16	24	32	40	48	56	64	72
9	9	18	27	36	45	54	63	72	81

Figure 3.4.7-4: Highlighted Even Rows.

So, statement #2 holds true for any (Column or Row) even number times any even or odd number. Combined is the geometric pattern in Figure 3.4.7-5.

	1	2	3	4	5	6	7	8	9
1	1	2	3	4	5	6	7	8	9
2	2	4	6	8	10	12	14	16	18
3	3	6	9	12	15	18	21	24	27
4	4	8	12	16	20	24	28	32	36
5	5	10	15	20	25	30	35	40	45
6	6	12	18	24	30	36	42	48	54
7	7	14	21	28	35	42	49	56	63
8	8	16	24	32	40	48	56	64	72
9	9	18	27	36	45	54	63	72	81

Figure 3.4.7-5: Highlight Even Columns and Rows.

So, statements #1, and #2 are highlighted in Figure 3.4.7-5, but the not highlighted values are multiplicand, multiplier, and products of statement #3, and that covers all 81 values.

Even times Even:	9 values and 4 rows	36
Even times Odd:	4 values and 5 rows	20
Odd times Odd:	5 values and 5 rows	25
For a total of		81

3.4.7 Summary.

The three statements are true and can be supported geometrically, which is something the Greeks used to prove multiplication division, and right triangles.

3.4.8: Pattern 8 –Mirror.

The Mirror pattern is one of the simplest patterns of the Multiplication table. Like the Column and Row of Pattern 6, it has its foundation in the Commutative Properties of multiplication.

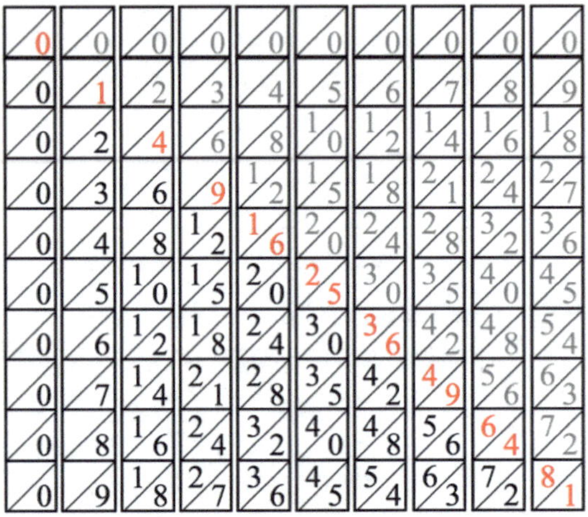

Figure 3.4.8-1: Notice the Squared Values were the Column and Row equal.

In Figure 3.4.8-1, a student only has to write the Multiplication table twice to recognize that any two numbers are repeated above-right and lower-left of the square values in red. The above-right is a mirror image of the lower-left, which is where the pattern name came from. In this case, to simplify seeing the pattern, the zero (0) row is restored to make a 10x10 square.

3.4.8.1 Upper and to the Left diagonals.

A few examples will fortify the idea of a mirror pattern:

0x9: Starting at the lower-left corner, one can see that 0x9 = 0; and at the upper-right corner that 9x0 = 0. Two properties of multiplication are clear, that the Commutative Properties of zero hold true.

Between these two products of zero (0) is a diagonal line of products that replicate each other as you move inward to the non-middle point. This non-middle point means that there is no single exclusive product, but two equal values.

0, 8, 14, 18, 20, 20, 18, 14, 8, 0.

The non-middle point is because this diagonal is based on nine (9), which is an odd number. This odd number can be any row multiplier or column multiplicand.

In Figure 3.4.8.2, you can see the left-to-right upward diagonal of the above 10 values.

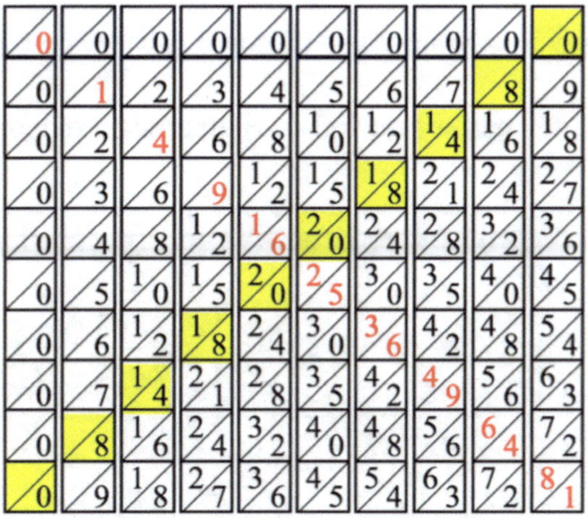

Figure 3.4.8-2: Mirror Pattern of 0x9.

0x8: The next value is up the zero-column and is matched by the 8x0 product where both are equal to zero (0). The diagonal products between them are 0, 7, 12, 15, 16, 15, 13, 7, 0.

In this case, because eight (8) is an even number, we have a middle value (16), which happens to be a square root. More about that later; however, it is clearly not the square of nine (9) or eight (8).

In Figure 3.4.8-3, you can see the left-to-right upward diagonal of the above 9 values.

Figure 3.4.8-3: Mirror Pattern of 0 x8.

0x7: this column and row value have an equal row, column value of 7x0, and product of zero (0). Where the diagonal products between them are 0, 6, 10, 12, 12, 10, 6, 0.

The non-middle point is because this diagonal is based on the seven (7) being an odd number.

In Figure 3.4.8-4, you can see the left-to-right upward diagonal of the above 8 values.

Figure 3.4.8-4: Mirror Pattern of 0x7.

0x6: this column and row value have an equal row, column value of 6x0, and product of zero (0). Where the diagonal products between them are 0, 5, 8, 9, 8, 5, 0.

This multiplication problem has a middle value because the diagonal is based on six (6) being an even number.

In Figure 3.4.8-5, you can see the left-to-right upward diagonal of the above 7 values.

Figure 3.4.8-5: Mirror Pattern of 0x6.

0x5: this column and row value have an equal row, column value of 5x0, and product of zero (0). Where the diagonal products between them are 0, 4, 6, 6, 4, 0.

The non-middle point is because this diagonal is based on five (5) being an odd number.

In Figure 3.4.8-6, you can see the left-to-right upward diagonal of the above 6 values.

Figure 3.4.8-6: Mirror Pattern of 0x5.

0x4: this column and row value have an equal row, column value of 4x0, and product of zero (0). Where the diagonal products between them are 0, 3, 4, 3, 0.

This multiplication problem has a middle value because the diagonal is based on the four (4) being an even number.

In Figure 3.4.8-7, you can see the left-to-right upward diagonal of the above 5 values.

Figure 3.4.8-7: Mirror Pattern of 0x4.

0x3: this column and row value have an equal row, column value of 3x0, and a product of zero (0). Where the diagonal products between them are 0, 2, 2, 0.

The non-middle point is because this diagonal is based on an odd number of three (3).

In Figure 3.4.8-8, you can see the left-to-right upward diagonal of the above 4 values.

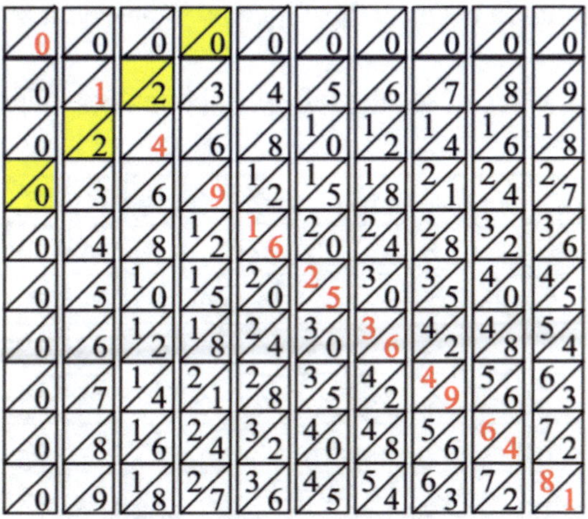

Figure 3.4.8-8: Mirror Pattern of 0x3.

0x2: this column and row value have an equal row, column value of 2x0, and product of zero (0). Where the diagonal products between them are 0, 1, 0.

This multiplication problem has a middle value because the diagonal is based on an even number of two (2).

In Figure 3.4.8-9, you can see the left-to-right upward diagonal of the above 3 values.

Figure 3.4.8-9: Mirror Pattern of 0x2.

"0x1: this column and row value have an equal row and column value of 1x0 with a product of zero (0). Thus the diagonal products between them are 0, 0.

The non-middle point is because this diagonal is based on an odd number of one (1).

In Figure 3.4.8-10, you can see the left-to-right upward diagonal of the above 2 values.

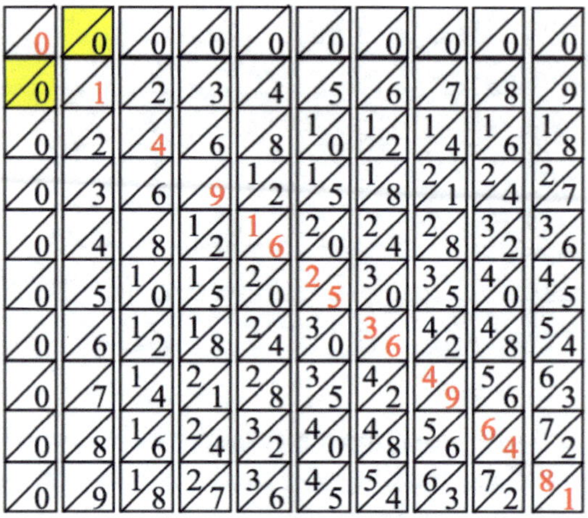

Figure 3.4.8-10: Mirror Pattern of 0x1.

0x0: this column and row value of 0x0, and row and column value of 0x0, and has no middle or non-middle value because there is no diagonal, and the value is a square of zero (0). As such, Figure 3.4.8-11 is one of the proofs that zero is neither odd nor even. It is also one of the reasons that Napier's bones or Tally-x sticks do not include the zero (0) row.

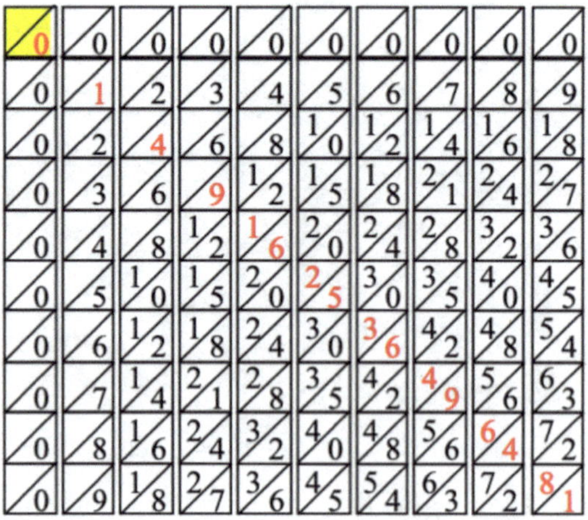

Figure 3.4.8.11: Mirror Pattern of 0x0.

There is not a table that shows the left-right upward diagonal, as this is a corner square of the upper left corner.

Note: that the number of values in the diagonal is one (1) more than the multiplier. In addition, note that the first/last value is zero (0), and the second/2nd from last is one (1) less than the multiplier.

3.4.8.2 Lower and to the Right Diagonals.

In 3.4.8.1, the left-to-right upward diagonals moved upward toward the top left corner. In 3.4.8.2, the same approach will be used to go to the lower right corner. In this case, nine (9) represents the ninth (9) row, and the multiplier is the column value.

9x1: this column and row value have an equal row, column value of 3x0, and product of zero (0). The diagonal products between them are 9, 16, 21, 24, 25, 24, 21, 16, 9.

The non-middle point is because this diagonal is based on one (1) being an odd number.

In Figure 3.4.8-12, you can see the left-to-right upward diagonal 9-values above.

Figure 3.4.8-12: Mirror Pattern of 9x1.

9x2: this column and row value have an equal row, and column value of 2x0, and product of zero (0). The diagonal products between them are 18, 24, 28, 30, 30, 28, 24, 18.

The middle point is because this diagonal is based on two (2) being an even multiplier number.

In Figure 3.4.8-13, you can see the left-to-right upward diagonal 8-values above.

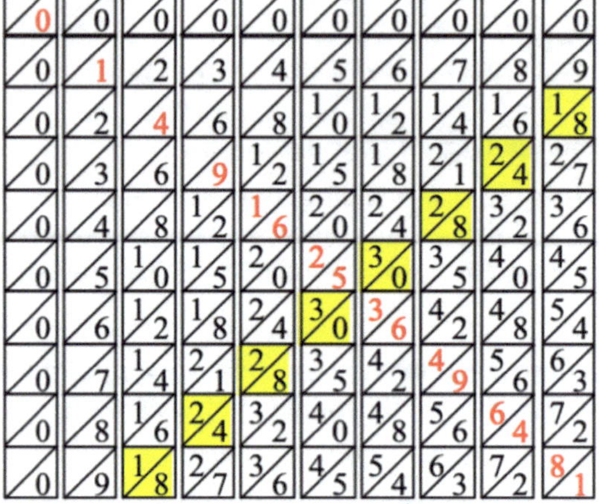

Figure 3.4.8-13: Mirror Pattern of 9x2.

9x3: this column and row value have an equal row, column value of 1x0, and product of zero (0). The diagonal products between them are 27, 32, 35, 36, 35, 32, 27.

The non-middle point is because this diagonal is based on three (3) being an odd number.

In Figure 3.4.8-14, you can see the left-to-right upward diagonal values 7-values above.

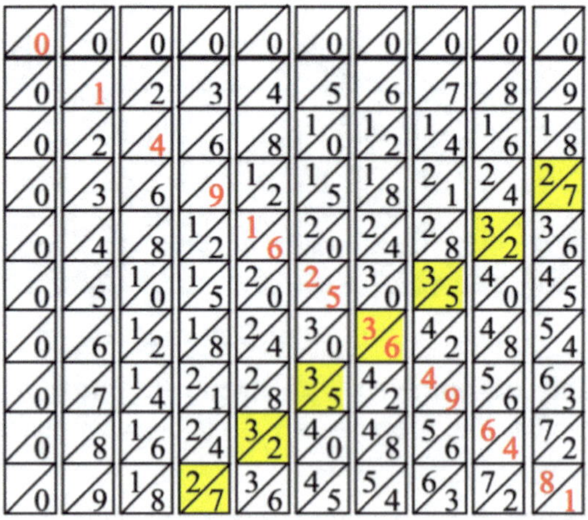

Figure 3.4.8-14 Mirror Pattern of 9x3.

9x4: this column and row value have an equal row, column value of 1x0, and product of zero (0). The diagonal products between them are 36, 40, 42, 42, 40, 36.

The middle point is because this diagonal is based on four (4) being an even multiplier number.

In Figure 3.4.8-15, you can see the left-to-right upward diagonal values 6-values above.

Figure 3.4.8-15: Mirror Pattern of 9x4.

9x5: this column and row value have an equal row, column value of 1x0, and product of zero (0). The diagonal products between them are 45, 48, 49, 48, 45.

The non-middle point is because this diagonal is based on five (5) being an odd number.

In Figure 3.4.8-16, you can see the left-to-right upward diagonal values 5-values above.

Figure 3.4.8-16: Mirror Pattern of 9x5.

9x6: this column and row value have an equal row, column value of 1x0, and product of zero (0). The diagonal products between them are 54, 56, 56, 54.

The middle point is because this diagonal is based on six (6) being an even multiplier number.

In Figure 3.4.8-17, you can see the left-to-right upward diagonal 4-values above.

Figure 3.4.8-17: Mirror Pattern of 9x6.

9x7: this column and row value have an equal row, column value of 1x0, and product of zero (0). The diagonal products between them are 63, 64, 63.

The non-middle point is because this diagonal is based on seven (7) being an odd number.

In Figure 3.4.8-18, you can see the left-to-right upward diagonal 3-values above.

0	0	0	0	0	0	0	0	0	0
0	1	2	3	4	5	6	7	8	9
0	2	4	6	8	1/0	1/2	1/4	1/6	1/8
0	3	6	9	1/2	1/5	1/8	2/1	2/4	2/7
0	4	8	1/2	1/6	2/0	2/4	2/8	3/2	3/6
0	5	1/0	1/5	2/0	2/5	3/0	3/5	4/0	4/5
0	6	1/2	1/8	2/4	3/0	3/6	4/2	4/8	5/4
0	7	1/4	2/1	2/8	3/5	4/2	4/9	5/6	6/3
0	8	1/6	2/4	3/2	4/0	4/8	5/6	6/4	7/2
0	9	1/8	2/7	3/6	4/5	5/4	6/3	7/2	8/1

Figure 3.4.8-18: Mirror Pattern of 9x7.

9x8: this column and row value have an equal row, column value of 1x0, and product of zero (0). The diagonal products between them are 72 and 72.

The middle point is because this diagonal is based on eight (8) being an even multiplier number.

In Figure 3.4.8-19, you can see the left-to-right upward diagonal 2-values above.

0	0	0	0	0	0	0	0	0	0
0	1	2	3	4	5	6	7	8	9
0	2	4	6	8	1/0	1/2	1/4	1/6	1/8
0	3	6	9	1/2	1/5	1/8	2/1	2/4	2/7
0	4	8	1/2	1/6	2/0	2/4	2/8	3/2	3/6
0	5	1/0	1/5	2/0	2/5	3/0	3/5	4/0	4/5
0	6	1/2	1/8	2/4	3/0	3/6	4/2	4/8	5/4
0	7	1/4	2/1	2/8	3/5	4/2	4/9	5/6	6/3
0	8	1/6	2/4	3/2	4/0	4/8	5/6	6/4	7/2
0	9	1/8	2/7	3/6	4/5	5/4	6/3	7/2	8/1

Figure 3.4.8-19: Mirror Pattern of 9x8.

9x9: this column and row value of 81 has no middle or non-middle value because there is no diagonal.

3.4.8.3 Mirror Pattern.

The above examples of section 3.4.8.1 are only what covers the upper triangle (blue) that duplicates the associated left triangle (blue) Figure 3.4.8-20.

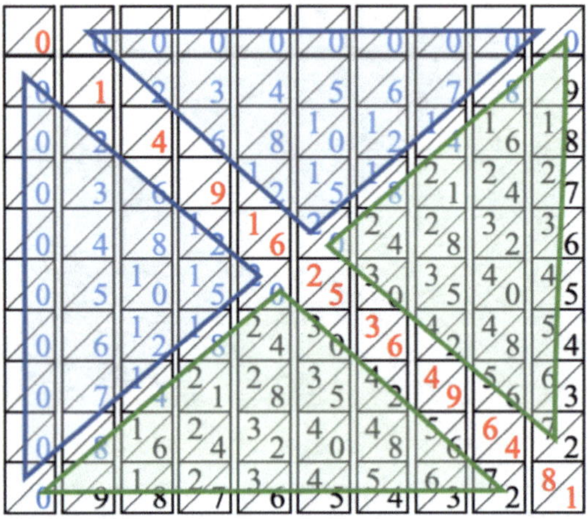

Figure 3.4.8-20: Mirror pattern of the Upper Triangle and the Left-side Triangle.

From the "blue" examples above, the first example could be transposed, and the multiplicand (column) value (0x9) becomes the multiplier (row) value, vice versa (9x0) or the "green" triangles as seen in section 3.4.8.2. This would then complete the "mirror pattern."

The only thing left is the squared values along the diagonal values in "red" font, which are the squares of the horizontal (row), and vertical (column) values. Therefore, the product of the same value, row, and column, is the square, *reference 2.10,* of the number.

3.4.8 Summary.

This pattern is typically seen by students in grade school as well where they are required to repeatedly create the multiplication table without mistakes. As a pattern, this results in few or no mistakes, but it is often forgotten over the years.

So, what is the Pattern here and in Figure 3.4.8-21 below?

	Squares									Non-Squares									
Upper Left	Odd Values Out from Center									Even Values Out from Center									
Counts from Center	-7	-5	-3	-1		-1	-3	-5	-7		-6	-4	-2	0		-2	-4	-6	
	Mixed Odd & Even Values									All Even Values									
0 x 0					0														
0 x 1														0	0				
0 x 2				0	1	0													
0 x 3													0	2	2	0			
0 x 4			0	3	4	3	0												
0 x 5												0	4	6	6	4	0		
0 x6		0	5	8	9	8	5	0											
0 x 7											0	6	10	12	12	10	6	0	
0 x 8	0	7	12	15	16	15	12	7	0										
Middle																			
0 x 9										0	8	14	18	20	20	18	14	8	0
Lower Right																			
9 x 1	9	16	21	24	25	24	21	16	9										
9 x 2											18	24	28	30	30	28	24	18	
9 x 3		27	32	35	36	35	32	27											
9 x 4												36	40	42	42	40	36		
9 x 5			45	48	49	48	45												
9 x 6													54	56	56	54			
9 x 7				63	64	63													
9 x 8														72	72				
9 x 8					81														

Figure 3.4.8-21: The Addition Behind the Mirror Pattern- Odds and Evens.

For the diagonals that pass through a square value, each value left-downward or right-upward from the square value changes by a sequence of odd decrements: 1, 3, 5, and 7. Thus, the first change is 1 less than the square, the second change is 3 less than the previous value, the fourth change is 5 less than the previous value, and then the fifth change is 7 less than the previous value. This applies equally whether above the main diagonal or below.

For the diagonals that do not pass through a square value, each value left-downward or right-upward from the square value changes by a sequence of even decrements: 0, 2, 4, and 6. Thus, the first change is 0 less than the complement value (i.e., they are the same value), the second change is 2 less than the previous value, the fourth change is 4 less than the previous value, and then the fifth change is 6 less than the previous value. This applies equally whether above the main diagonal or below.

3.4.9: Pattern 9 – Complementary Diagonals.

Just as you can find patterns in the multiplication table in Columns and Rows, you can also find patterns in the diagonal through the square values of the integers, 0 to 9. Similar, to the Mirror (Pattern 8) method that folds top-right across the squared values diagonally to the bottom-left.

3.4.9.1 Perfect Squares Diagonally.

In this case, to make it easier to visualize, we are going to reintroduce the multiplication row for zero (0). The first figure (Figure 3.4.9-1) shows squared values in green boxes where the multiplicand (row values) and multiplier (column values) are equal. For instance, 5x5 equals 25, as shown in the sixth row down and the sixth column from the left.

The diagonal green lines represent the 'complementary diagonal' solutions, mirrored above and below the perfect squared values. The diagonal line starts with a row value and travels up to its equivalent column value. Note: this only applies to even row values, which are also displayed in green text.

	0	1	2	3	4	5	6	7	8	9
0	0	0	0	0	0	0	0	0	0	0
1	0	1	2	3	4	5	6	7	8	9
2	0	2	4	6	8	10	12	14	16	18
3	0	3	6	9	12	15	18	21	24	27
4	0	4	8	12	16	20	24	28	32	36
5	0	5	10	15	20	25	30	35	40	45
6	0	6	12	18	24	30	36	42	48	54
7	0	7	14	21	28	35	42	49	56	63
8	0	8	16	24	32	40	48	56	64	72
9	0	9	18	27	36	45	54	63	72	81

Figure 3.4.9-1: Complementary Diagonals – Perfect Squares.

If we stopped here, then the geometric pattern would be no different than that of the Mirror pattern. However, Figure 3.4.9-2 shows us something mathematically different.

The blue values across the columns at the top are built outward from the zero (0) value which is the perfect square of the equal row, and column values, 0 to 9, in Figure 3.4.9.1-1 above. The multiplication and answer show the perfect square on a green background, just as in Figure 3.4.9-1 above. The blue values across the top will be explained later; however, let me finish the "layout" of the figure. The top five (5) rows are labeled "Even Row Values", which are equivalent to the green values in Figure 3.4.9.1-1 above. Look carefully, and you will see that the multiplicand (first value in the multiplication) is also in bold text, and represents the Row Value.

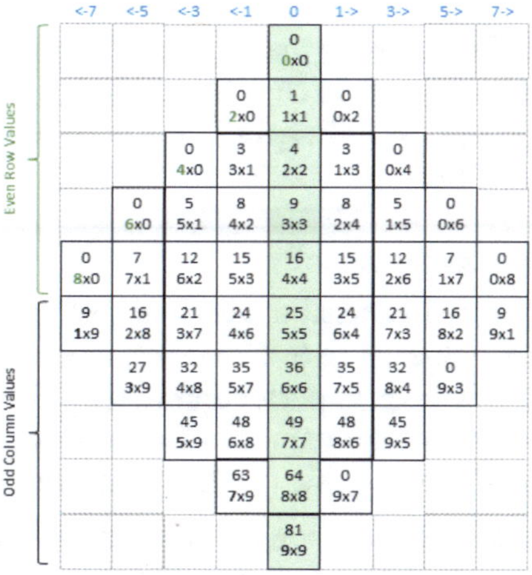

Figure 3.4.9-2: Complementary Diagonals - Layout.

You will also note that for 0 time 0 is the top green square and the top-left green square in Figure 3.4.9-2. There are no values up or down from the diagonal line, as such the figure has no values to the left or right of the squared zero (0) value.

Using the same approach, 1x1 has one value up, and down on the diagonal above, and so too in the figure will you find one value to the left, and right of the one squared value. This approach continues down the above-even row values to the eighth (8) value. For the 8th value, you will find four (4) values on either side of 4 squared or 16 answer squares.

The value in each square, above the multiplication, is the answer. I.e., 2x2 is 4. In the figure, there are two values on either side of the perfect square value of 2x2, with answers of 3, and 0. The three (3) value is one (1) less than the squared value (4), and the zero (0) value is 3 less than the 3x1 (or 1x3) value. Notice that these differences are listed in blue for each applicable column., and most importantly it is the same pattern of difference for each column to either side of the squared value, always the odd value from 0 to 8, the complementary odd value of each even row value.

The second part of Figure 3.4.9-2 is the lower part or Odd Column Values. As can be seen in Figure 3.4.9-1, when we reach the last even row value, the diagonals wrap around the bottom and reflect the first 9x1 value or the first odd column value. It repeats that across all the odd column values up to nine

(9). So, for the first square in the lower section, the multiplicand is the odd column value for that squared value. I.e., 5x5, 6x6, … 9x9., and again, outward from the squared values each square's answer is reduced by the column blue values.

The multiplicand value is one more than the number of values in the same row. I.e., for the multiplicand of six (6), the number of values should be eight (8), which is: 6x0, 5x1, 4x2, 3x3, 2x4, 1x5, and 0x6. Also, the squared integer is half the row value.

3.4.9.2 In-between Squares Diagonally.

Now there are also Mirrored values between each of the squared values that come from the odd row values: 1 to 9. However, nine (9) is like the squared diagonal, in that it divides the grid perfecting in two. However, in this case, there is no observed Mirror of the values in the lower-right to the values in the upper-left if folded along the nine (9) diagonal value.

In Figure 3.4.9.3, we have gray diagonal lines that start with each row value and travel up to its equivalent column value. This only occurs for the odd row values, which are in black text, because they cross between the perfect square blocks (green).

	0	1	2	3	4	5	6	7	8	9
0	**0**	0	0	0	0	0	0	0	0	0
1	0	**1**	2	3	4	5	6	7	8	9
2	0	2	**4**	6	8	10	12	14	16	18
3	0	3	6	**9**	12	15	18	21	24	27
4	0	4	8	12	**16**	20	24	28	32	36
5	0	5	10	15	20	**25**	30	35	40	45
6	0	6	12	18	24	30	**36**	42	48	54
7	0	7	14	21	28	35	42	**49**	56	63
8	0	8	16	24	32	40	48	56	**64**	72
9	0	9	18	27	36	45	54	63	72	**81**

Figure 3.4.9-3: Complementary Diagonals – In-Between Squares.

In Figure 3.4.9-3, we see something mathematically different from Figure 3.4.9-2 in that there is no center green box column. We are addressing the odd row values, and therefore we have an even count of values along each grey line.

The blue values across the columns at the top are built outwardly paired, 0 to 0, 2 to 2, 4 to 4, and 6 to 6, from the center as in Figure 3.4.9-3 above. Each is an even value. The blue values across the top, as explained in 3.4.9.1 represent the difference between the column to the right, and the current column: 0, 2, 4, and 6.

In Figure 3.4.9-4, the top five (5) rows are labeled "Odd Row Values", which are equivalent to the values in Figure 3.4.9-3 above. Look carefully, and you will see that the multiplicand (first value in the multiplication) is also in black/bold text, and represents the Row Value.

Note that just as explained in Section 3.4.9.1, there is no value to either side of the 0s columns, as such the figure has no values to the left or right of the two zero (0) column values.

Using the same approach, 2x2 has one value up, and down on the diagonal figure above, and so too in the figure will you find one value to the left, and right of the one squared value. This approach continues down the above odd row values to the ninth (9) diagonal line. For the 9th value, you will find four (4) values on either side of two 4x5 and 5x4 blocks.

As in Section 3.4.9.1, the value in each square, above the multiplication, is the answer. I.e., 4x3 or 3x4 is 12. In the figure, there are three values on either side of the 4x3 or 3x4 blocks, with answers of 10, 6, and 0. The ten (10) value is two (2) less than the 4x3 or 3x4 blocks, and the six (6) values are 4 less than the 5x2 (or 2x5) value. Notice that these differences are listed in blue for each applicable column., and most importantly it is the same pattern of difference for each column to either side of the 4x3 or 3x4 blocks.

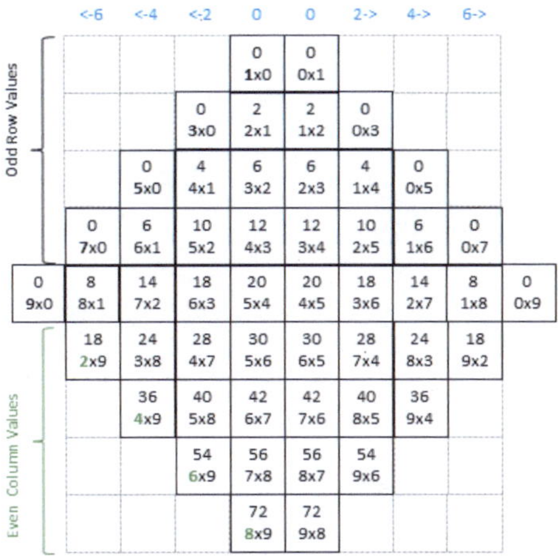

Figure 3.4.9-4: Complementary Diagonals – Layout.

The second part of Figure 3.4.9-4 is the lower part of Even Column Values and is the even version of the second part of Section 3.4.9.1. As we reach the last even row value, the diagonals wrap around the bottom and reflect the first 9x2 value or the first odd column value. It repeats that across all the even column values up to eight (8). So, for the first square in the lower section, the multiplicand is the even column value for 0 to 0 columns., and again, moving outward each square's answer is reduced by the column blue values.

Here the multiplicand value is also one more than the number of values in the same row. I.e., for the multiplicand of seven (7) the number of values should be eight (8), which is: 7x0, 6x1, 5x2, 4x3, 3x4, 2x5, 1x6, and 0x7. As this grouping is in-between squared values, it should not, and does not have an integer mid-value or squared mid-value.

3.4.9.3 Summary.

In the Figures above, the values on either side of the squared values (Section 3.4.9.1), and the in-between squared values (Section 3.4.9.2) are due to, and also proof of the Commutative Properties of Multiplication, which states the total solution of two multiplied values is the same answer no matter which is the multiplier, and which is the multiplicand. So, if there is a pattern out from each answer on either side of a square or in-between squared value, it will be the same on either side, start small, and get bigger till the solution is zero.

3.4.10: Pattern 10 – Squares.

Just as in the previous Complementary Diagonal of Pattern 9 that runs top-right down to bottom-left, there is a complementary diagonal that runs top-left down to bottom-right. This is the Square pattern where the maximum length happens to be the squares of the common multiplicand and multiplier. I.e., 1x1 = 1. In Figure 3.4.10.1-1, the square product is highlighted in yellow where the multiplicand is the column value, and the multiplier is the row value.

An interesting fact is that there are no squared 1's place values of: 2, 3, 7, or 8. Instead, the 1s place values have a pattern of: 0, 1, 4, 9, 6, 5, 6, 9, 4, as seen in Figure 3.4.10.1-1. The first part of the sequence starts with 0 followed by 1, 4, 9, 6 to 5, and the second part is reversed from 5 and followed by 6, 9, 4, 1, and back to 0. Associating a value with something else is a memorization method. So the 1, 4, 9, 6 values can be historically associated with the year 1496. This is the year Christopher Columbus sailed back to Spain to end his visit to the Americas, on the 5th of March 1496.

3.4.10.1 Squares.

In this section, we are focused on just one diagonal, which is also the longest diagonal, it shows the square of a of the common multiplicand and multiplier, shown in red. Here we will ignore the zero (0) squared, and start with the one (1) squared. See *reference* 3.4.

In Figure 3.4.10.1, the column, and row values are the same, so it is referenced with n in the figure.

Figure 3.4.10 -1: Math Pattern of a Squared Value n = 1.

In Figure 3.4.10-1 shows the result of multiplying 1 x1 = 1. This can be shown geometrically as a unit square of 1, with dimensions of 1 by 1. This will be repeated in the following n values.

Figure 3.4.10-2: Math Pattern of a Squared Value n = 2.

In Figure 3.4.10-2, we see the previous 1 x1 area value in gray, but the next unit larger is in yellow. This can be written as 1 (the previous squared unit) + 3 (the newest part of this square unit) = 4 squares, or 22 (n = 2). Where n = 2 is the column, and row value.

Although this is not obvious for n = 1, it can be seen in n = 2, 3, 4, 5, 6, 7, 8, and 9 that:

- The squared value, the largest of the geometric square, is always found in the lower right corner and is highlighted in red. This is also true of the previous squared units which are highlighted in red.
- The top yellow square in the column and the left yellow square in the row are both values of the column, and row to be squared, n.
- The odd number difference of the current square from the previous square results in n + n -1. This is a simple ruling that 1) the addition of two or more even numbers is an even number and 2) an even number less one is always odd. In this case, 2 + 2 – 1 is 3, which is the odd difference between 1 and 4. It is also the number of yellow squares.

Figure 3.4.10-3: Math Pattern of a Squared Value n = 3.

In Figure 3.4.10-3, we see the previous 2 x2, and 1 x1 area values in gray, but the next unit larger 3 x3 = 9 is in yellow. This can be written as 4 (the previous squared unit) + 5 (the newest part of this square unit) = 9 squares, or 32 (n = 3).

In this case, 3 + 3 – 1 is 5 which is the odd difference between 4 and 9, which can be confirmed by counting the yellow squares.

Figure 3.4.10-4: Math Pattern of a Squared Value n = 4.

In Figure 3.4.10-4, we see the previous area values in gray, but the next unit larger 4 x4 = 16 is in yellow. This can be written as 9 (the previous squared unit) + 7 (the newest part of this square unit) = 16 squares, or 42 (n = 4).

In this case, 4 + 4 – 1 is 7 which is the odd difference between 9, and 16, which can be confirmed by counting the yellow squares.

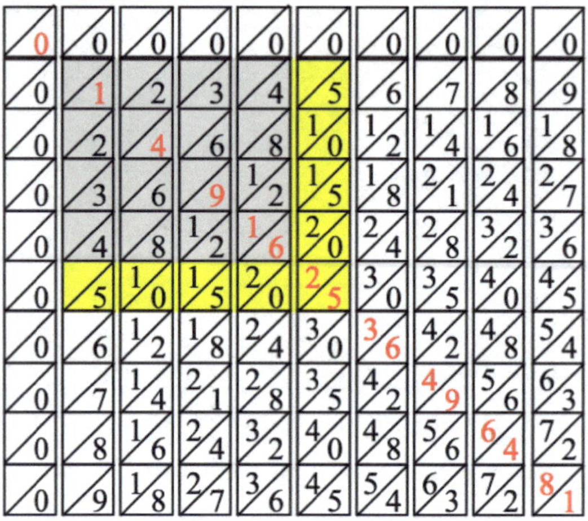

Figure 3.4.10-5: Math Pattern of a Squared Value n = 5.

In Figure 3.4.10-5, we see the previous 4 x4 area value in gray, but the next unit larger 5 x5 = 25 is in yellow. This can be written as 16 (the previous squared unit) + 9 (the newest part of this square unit) = 25 squares, or 52 (n = 5).

In this case, 5 + 5 – 1 is 9 which is the odd difference between 16 and 25, which can be confirmed by counting the yellow squares.

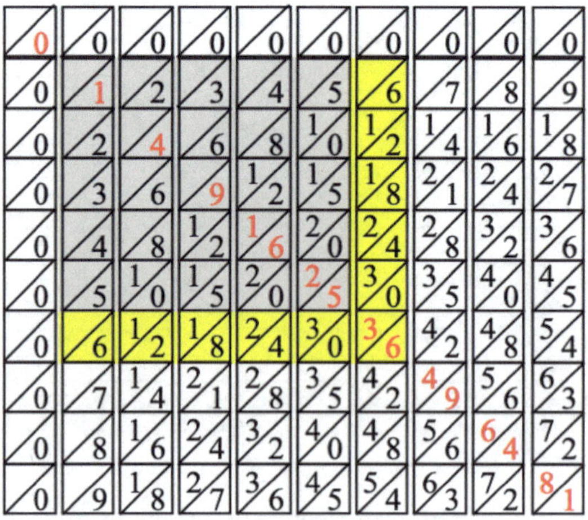

Figure 3.4.10-6: Math Pattern of a Squared Value n = 6.

In Figure 3.4.10-6, we see the previous 5 x5 area value in gray, but the next unit larger 6 x6 = 36 is in yellow. This can be written as 25 (the previous squared unit) + 11 (the newest part of this square unit) = 36 squares, or 62 (n = 6).

In this case, 6 + 6 – 1 is 11 which is the odd difference between 25 and 36, which can be confirmed by counting the yellow squares.

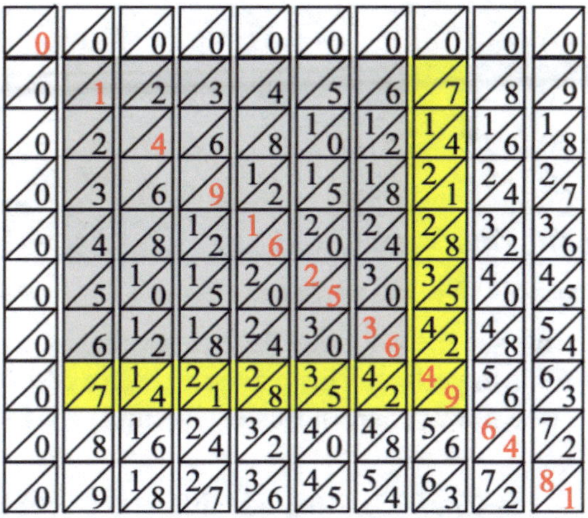

Figure 3.4.10-7: Math Pattern of a Squared Value n = 7.

In Figure 3.4.10-7, we see the previous 6 x6 area value in gray, but the next unit larger 7 x7 = 49 is in yellow. This can be written as 36 (the previous squared unit) + 13 (the newest part of this square unit) = 49 squares, or 72 (n = 7).

In this case, 7 + 7 – 1 is 13 which is the odd difference between 36 and 49, which can be confirmed by counting the yellow squares.

Figure 3.4.10-8: Math Pattern of a Squared Value n = 8.

In Figure 3.4.10-8, we see the previous 7 x7 area value in gray, but the next unit larger 8 x8 = 64 is in yellow. This can be written as 49 (the previous squared unit) + 15 (the newest part of this square unit) = 64 squares, or 82 (n = 8).

In this case, 8 + 8 – 1 is 15 which is the odd difference between 49 and 64, which can be confirmed by counting the yellow squares.

Figure 3.4.10-9: Math Pattern of a Squared Value n = 9.

In Figure 3.4.10-9, we see the previous 8 x8 area value in gray, but the next unit larger 9 x9 = 81 is in yellow. This can be written as 64 (the previous squared unit) + 17 (the newest part of this square unit) = 81 squares, or 92 (n = 9).

In this case, 9 + 9 – 1 is 17 which is the odd difference between 64 and 81, which can be confirmed by counting the yellow squares.

3.4.10.2 Pattern of Squares (Advanced).

If we create a table of the indexed value (i), the indexed squared (i^2), and the delta (?) as in Table 3.4.10-1. This delta, or difference value from the previous i and current i values. This shows a developing pattern between the index (i) and the delta (Δ).

Whether one takes the sum of the common value, and its previous value (Δ+) or the difference of the current squared value, and its previous squared value (Δ-), you get the same answer. The answer is a sequence of the odd numbers: 1, 3, 5, 7, …

Index	i^2	Δi	∑Δ
0	0	-	-
1	1	1	1
2	4	3	4
3	9	5	9
4	16	7	16
5	25	9	25
6	36	11	36
7	49	13	49
8	64	15	61
9	81	17	81

Table 3.4.10-1: Math Pattern of a Squared Value.

Therefore, using the delta sum (Δ+) we could postulate a test where the previous index is 10, and the current index is 11, then 10 squared (10^2) is 100. Now for the mathematical pattern: 10 (the first index) plus 11 (the second index) adds up to 21. Add this to the first index square and you get 100+21 or 121.

Is the pattern successful? Yes, because 11 squared (11^2) is 121. more than the 100, or 121. You should try this yourself for index values of 12 and 13.

Solutions

9 and 10:	$9^2 = 81 \Rightarrow$	$9 + 10 = 19$	Therefore, $81 + 19 = 100$	$10^2 = 100$
11 and 12:	$11^2 = 121 \Rightarrow$	$11 + 12 = 23$	Therefore, $121 + 23 = 144$	$12^2 = 144$
12 and 13:	$12^2 = 144 \Rightarrow$	$12 + 13 = 25$	Therefore, $144 + 25 = 169$	$13^2 = 169$
13 and 14:	$13^2 = 169 \Rightarrow$	$13 + 14 = 27$	Therefore, $169 + 27 = 196$	$14^2 = 196$
14 and 15:	$14^2 = 196 \Rightarrow$	$14 + 15 = 29$	Therefore, $196 + 29 = 225$	$15^2 = 225$
20 and 21:	$20^2 = 400 \Rightarrow$	$20 + 21 = 41$	Therefore, $400 + 41 = 441$	$21^2 = 441$

If you know a square near your value, i.e., $10^2 = 100$; add as above to get the next value. But, you can also get the value below it by subtracting!

15 and 14: $14^2 = 225$ ⇨ $14 + 15 = 29$

Therefore, $225 - 29 = 225$ => $14^2 = 196$

3.4.10.3 Summary.

Indeed, discovering these intricate patterns in math is like unraveling hidden treasures. It's where logic meets creativity, and every revelation brings a sense of satisfaction and wonder.

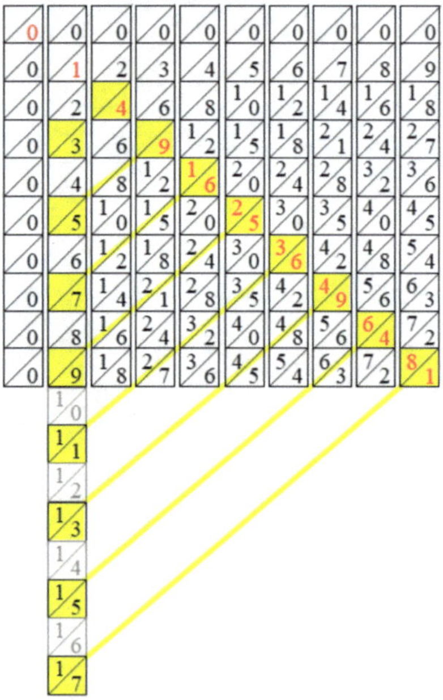

Figure 3.4.10-10: Diagonal Pattern between squared values.

3.4.11: Pattern 11 – Non-Square Diagonals.

So, we have seen that the diagonal of the squares has a pattern. The next logical question is whether the other left-to-right downward diagonals will also have a pattern. We will use the diagonal to the left of the squares, 0 to 72 to show the even start values, and 0 to 63 to show the odd start values. See Figure 3.4.11-1 below. Note, that the Mirror image diagonal is the same just to the right of the squares, so we will not cover the right side.

In Figure 3.4.11.1, the column value, and the row value are not the same, so we will them separately with c (column), and r (row).

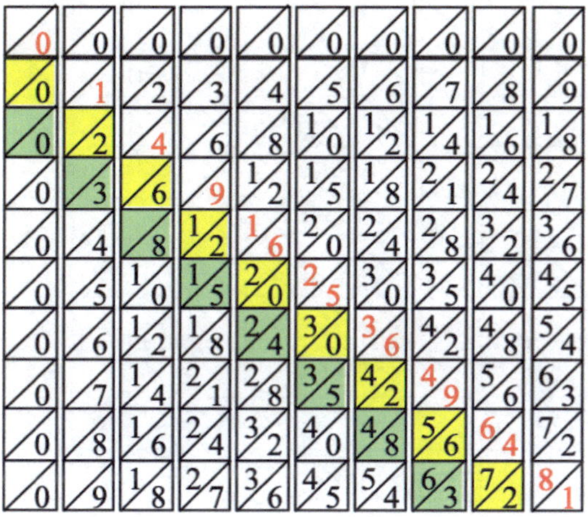

Figure 3.4.11-1: Left-to-right Downward Diagonals – c = 2 and c = 3 Start Values.

In Table 3.4.11-1, we see that for any even value starting (shown in red) from 2, the left-right downward numbers are all even in the table's first row. In fact, the differences (Δ) are also even in the table's second row. You also see that for any odd value starting (shown in green) from 3, the values are an odd-even repeating pattern in the table's first row. But, the second row is all incrementally the odd values that start with the odd.

Note: The 2 and 3 are the top non-zero value that indicates the row multiplication value.

The second row is plus 2 of the first value on the left, which is the red value. Thereby keeping the left side, even numbers, and the right side, odd numbers. i.e., 5 =(2) = 7, the next odd value.

The first row is the value below it and all numbers are to the left. I.e., first row value 20 is the sum of (2+4) + (6+8) or 6 + 14 = 20. This is the case for all such Left-to-right Patterns Tables below.

Even Value Start

0	2	6	12	20	30	42	56	72
	2	4	6	8	10	12	14	16

Odd Value Start

0	3	8	15	24	35	48	63
	3	5	7	9	11	13	15

Table 3.4.11-1: Math Pattern of Left-to-right Downward Diagonals – c = 2 and 3.

In Figure 3.4.11-2, the next set of even (4), and odd (5) start values from Figure 3.4.10.2-1 above also have a pattern. These diagonals are 0 to 54 to show the even start values, and 0 to 45 to show the odd Start values.

Figure 3.4.11-2: Left-to-right Downward Diagonals – c = 4 and c = 5 Start Values.

In Table 3.4.11-2, we see that for any even value starting (shown in red) from 4, then left-right downward numbers are all even in the table's first row. Differences (Δ) are also even in the table's second row. You also see that for any odd value starting from 5, the values are an odd-even repeating pattern in the table's first row. But, the second row is all incrementally the odd values that start with the odd value.

Even Value Start

0	4	10	18	28	40	54
	4	6	8	10	12	14

Odd Value Start

0	5	12	21	32	45
	5	7	9	11	13

Table 3.4.11-2: Math Pattern of Left-to-right Downward Diagonals – c = 4 and 5.

In Figure 3.4.11-3, the next set of even (6), and odd (7) starting values from Figure 3.4.10.2-2 above also have a pattern. These diagonals are 0 to 36 to show the even start values, and 0 to 27 to show the odd start values.

Figure 3.4.11-3: Left-to-right Downward Diagonals – c = 6 and c = 7 Start Values.

In Table 3.4.11-3, we see that for any even value starting (shown in red) from 6, the left-right downward numbers are all even in the table's first row. The differences (Δ) are also even in the table's second row. You also see that for any odd value starting from 7, the values are an odd-even repeating pattern in the table's first row. But, the second row is all incrementally the odd values that start with an odd value.

Even Value Start

0	6	14	24	36
	6	8	10	12

Odd Value Start

0	7	16	27
	7	9	11

Table 3.4.11-3: Math Pattern of Left-to-right Downward Diagonals – c = 6 and 7.

In Figure 3.4.11-4, the next set of even (8), and odd (9) start values from Figure 3.4.11-3 above also have a pattern. These diagonals are 0 to 18 to show the even start values and 0 to 9 to show the odd start values.

Figure 3.4.11-4: Left-to-right Downward Diagonals – c = 8 and c = 9 Start Values.

In Table 3.4.11-4, we see that for any even value starting (shown in red) from 8, the left-right downward numbers are all even in the table's first row. The differences (Δ) are also even in the table's second row. You also see that for any odd value starting from 9, the values are an odd-even repeating pattern in the table's first row. But, the second row is all incrementally the odd values that start with the odd value.

Even Value Start

0	8	18
	8	10

Odd Value Start

0	9
	9

Table 3.4.11-4: Math Pattern of Left-to-right Downward – c = 8 and 9.

Also note as in all the above tables that from 3.4.10.1-1 down to 3.4.10.1-4 lose the left and right blocks, by retaining the zero (0) first block on the left.

3.4.11.1 Summary

In Table 3.4.11-5 we see the pattern in that the delta (Δ) is the same in the Left-Right upward diagonals. For example, the middle column of the Even Value start is all 8s. The same is true for the odd value start, where they are all the same downward. For example, the center is all 9s down.

Even Value Start

0	2	6	12	20	30	42	56	72
	2	4	6	8	10	12	14	16

0	4	10	18	28	40	54
	4	6	8	10	12	14

0	6	14	24	36
	6	8	10	12

0	8	18
	8	10

Odd Value Start

0	3	8	15	24	35	48	63
	3	5	7	9	11	13	15

0	5	12	21	32	45
	5	7	9	11	13

0	7	16	27
	7	9	11

0	9
	9

Table 3.4.11-5 Math Pattern of Left-to-right Downward diagonals.

Uncommon is the pattern of the other left-right downward diagonals on either side of the square diagonal. From Figure 3.4.11-6 the diagonal patterns have an odd, and even bias. However, the algebraic difference is there is no difference beginning with an odd or even value.

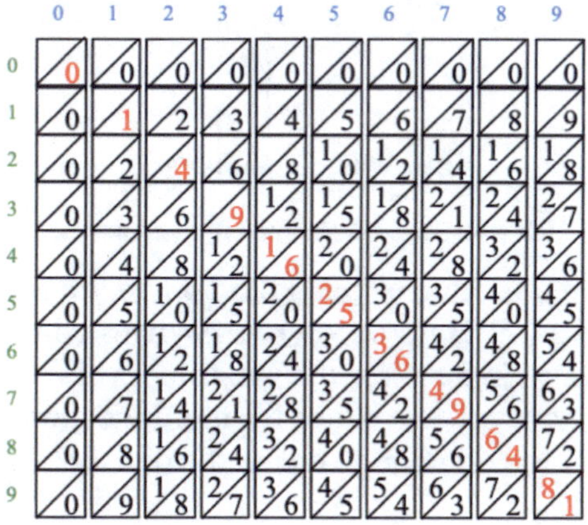

Figure 3.4.11-5: Summary.

3.4.11.2 Summary (Advanced)

Using Algebra this can be seen based on the starting value, c, and r. c + r +1 which are the blue numbers In Figure 3.4.10-6 above. This can also, be explained algebraically that if the prior squared value cxr, the next squared value would be (c+1) (r+1). The difference between them would be written:

$$(c+1)(r+1) - cxr = (cr + c + r + 1) - cr = c + r + 1 \qquad \text{Equation 3.4.11-1.}$$

3.4.12: Pattern 12 –Value Counting.

Just as in the previous Complementary Diagonal of Pattern 9 that runs top-right down to bottom-left, there is a complementary diagonal that runs top-left down to bottom-right. This is the value pattern where the maximum count happens to be the bottom square of the common multiplicand and multiplier row value. I.e., 1 x9 = 1.

3.4.12.1: 1 and 2 digit counting.

In Figure 3.4.12-1 the value count is highlighted in yellow as the multiplicand times the multiplier of row 1. In this Figure, the multiplier of 1, the top row, only has one digit across the 0 to 9 sticks. An example of the 1- and 9-stick is shown in Figure 3.4.12-1 in yellow. Here there is only 1 value count in the 1-stick, but there is a 2 value count for the 9-stick.

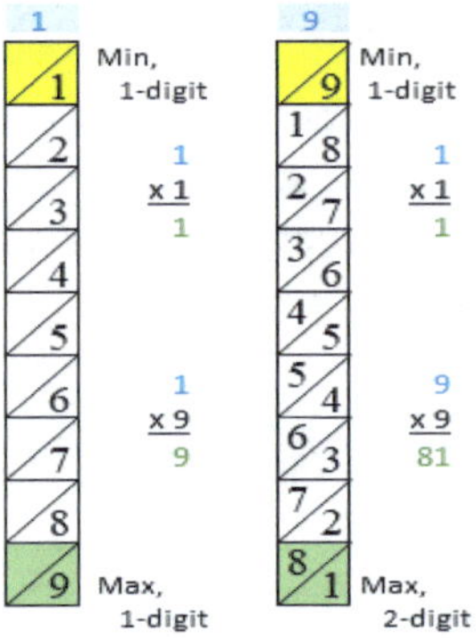

Figure 3.4.12-1: 1 and 2 digit counting.

Therefore, the key to the value count is the 1st and 9th rows. The 1st row is simply 1 for all sticks 0-9. But, the 9th row changes with the 2-stick with a value count of 2. For a single stick, the maximum value count starts at 1. But, what happens when you use two sticks for multiplication to create the multiplicand, and for the division to create the divisor?

3.4.12.2: 2 and 3 digit counting.

But, what happens when you use these individual sticks to gather the value count for multiplication, to create the multiplicand, and for division, to create the divisor? In the figure, the value count is still highlighted in yellow as the multiplicand times the multiplier of row 1. In this figure, the multiplier of 1, the top row, only has one digit for each stick across the 0 to 9 sticks. An example of the 11- and 99-stick is shown in Figure 3.4.12-2 in yellow. Here there are 2 value counts for the 11-stick and for the 99-stick.

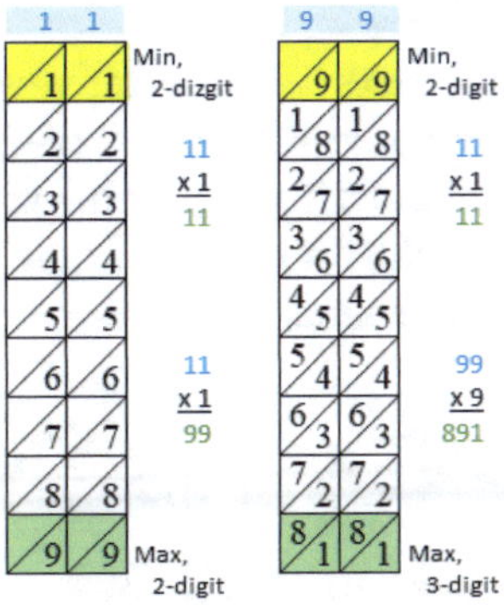

Figure 3.4.12-2: 2 and 3 digit counting.

Now in this figure of 2 sticks each, the 9th row changes the maximum value counts starting at the 11-stick setup it shows a value of 99 or a count of 2. But, for the 99-stick you get an 891 value or a count of 3. This will be found useful for multiplication, to create the multiplicand, and for division, to create the divisor.

But, the last two figures were for single digit multipliers, or the row numbers. But, for 2-digit multipliers, you can increase the value count by one. Similar to Figure 3.4.12-2, the only difference is that Figure 3.4.12-3 has two-digit multiplier examples to the right of the sticks.

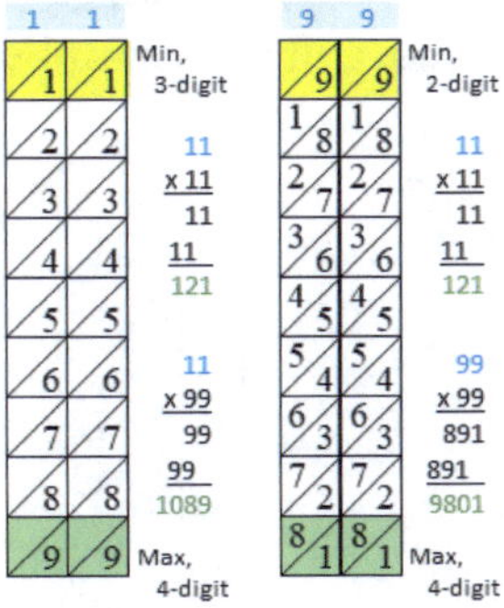

Figure 3.4.12-3: 2 to 3 digit counting.

You can see that for each extra value count of the multiplier, you add one more maximum value count.

3.4.12.3: 3 and 4 digit counting.

Let us confirm this idea in the next two figures, where we first increase the multiplicand by a count of one over Figure 3.4.12-2 and keep the multiplier to a single digit. So, In Figure 3.4.12-4, the value count is highlighted in yellow, now instead of two sticks for each example we have three, giving us the example of 111- and the 999-stick sets below.

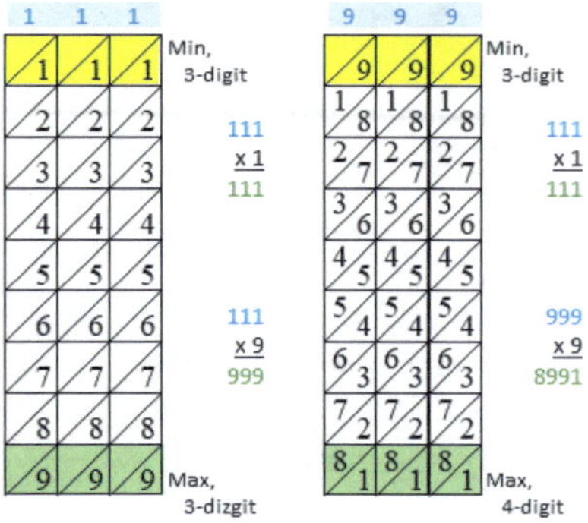

Figure 3.4.12-4: 3 to 4 digit counting.

You can see that for each extra value count of the multiplicand, you add one more maximum value count. So, in this case, the top value count is 3, but the maximum bottom value of 8991 has a value count of 4.

Ok, now let's keep Figure 3.4.12-4, but increase the multiplier count to 2. So, in Figure 3.4.12-5, the value count is highlighted in yellow where the multiplicand is the column value, and the multiplier is the row value.

Figure 3.4.12-5: 3 to 4 digit counting.

You can see that for each extra value count of the multiplier, you add one more maximum value count, and the practical application of this pattern is found in Chapter 6: Decimal Multiplication and Division.

3.4.12.4: Division.

For a preliminary practical application, Figure 3.4.12-6 shows a division problem where the divisor has a value count of 1 and the dividend has a value count of 2. The issue is that the 2 in the dividend is larger than the divisor and you have to use the maximum count of 2.

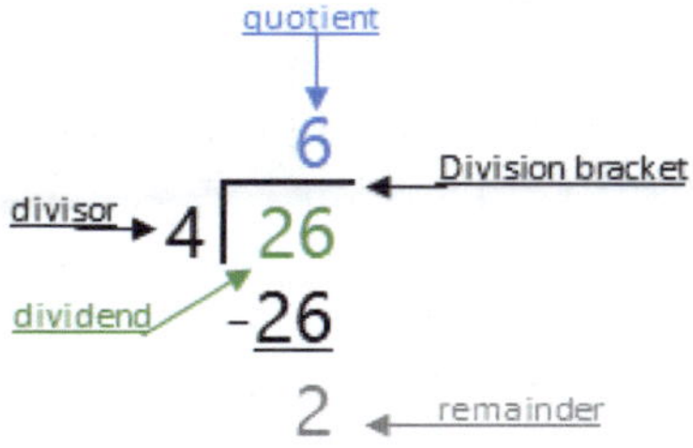

Figure 3.4.12-6: Division.

3.4.12 Summary.

The understanding here is that the value count of:

1. *Multiplication*: The value count of the multiplicand is the baseline count, but this can be increased by any extra counts in the multiplier, up to a maximum value of the baseline count plus one. For example, Figure 3.4.12-5 has a baseline count of 3 and a maximum count of 4.

2. *Division*: the value count of the multiplier is either the baseline count or the maximum count is the baseline plus one. For example, Figure 3.4.12-6 has a baseline count of 1 and a maximum count of 2. In this case, division requires the use of the maximum count to work. See Chapter 6.

3.4.13: Pattern 13 – Multiplication Geometric Patterns.

In this pattern, which is similar to Pattern 3.4.8: Mirror, we have the position pattern of odds and even in the 10x10 matrix.

3.4.13.1: Multiplication Pattern for 0.

This observation pertains to a vertical alignment over the column labeled "0", which might imply that zero (0) is an even number. However, according to mathematical principles, zero is classified as neither even nor odd.

0										
	0	1	2	3	4	5	6	7	8	9
	10	11	12	13	14	15	16	17	18	19
	20	21	22	23	24	25	26	27	28	29
	30	31	32	33	34	35	36	37	38	39
	40	41	42	43	44	45	46	47	48	49
	50	51	52	53	54	55	56	57	58	59
	60	61	62	63	64	65	66	67	68	69
	70	71	72	73	74	75	76	77	78	79
	80	81	82	83	84	85	86	87	88	89
	90	91	92	93	94	95	96	97	98	99

Figure 3.4.13-1: Digital Pattern of 0.

Mathematical *parity* is a rule that up all integers into two categories: even and odd numbers. This is done simply by asking whether the number divisible by 2, i.e., is the answer without a remainder or fractional component. If yes, then it is even. If no, it is odd by default.

It would imply that zero (0) is an even number in that it classifies as an integer, and when divided by 2 it results in 0 and has no remainder.

This can be seen in the first column of Figure 3.4.13-1 above. Where the 1s place value is always 0 and is divisible by 2, i.e., 10 /2 = 5, 20 /2 = 10, … 90 /2 = 45, with none having a remainder.

But, 0 added to 0 is still zero. So there is no movement in the Pattern of 0, and the figure remains empty.

3.4.13.2: Multiplication Pattern for 1.

In Figure 3.4.13-2 we have a 10-digit column and row of all numbers starting with zero (0).

1	0	1	2	3	4	5	6	7	8	9
	10	11	12	13	14	15	16	17	18	19
	20	21	22	23	24	25	26	27	28	29
	30	31	32	33	34	35	36	37	38	39
	40	41	42	43	44	45	46	47	48	49
	50	51	52	53	54	55	56	57	58	59
	60	61	62	63	64	65	66	67	68	69
	70	71	72	73	74	75	76	77	78	79
	80	81	82	83	84	85	86	87	88	89
	90	91	92	93	94	95	96	97	98	99

Figure 3.4.13-2: Digital Pattern of 1.

Starting with 1 and adding 1 sequentially, you get every number following 1.

3.4.13.3: Multiplication Pattern for 2.

First, make a copy of the multiplication table, and cut the columns into strips. Layout out each column on a table or tape them to a whiteboard.

2	0	1	2	3	4	5	6	7	8	9
	10	11	12	13	14	15	16	17	18	19
	20	21	22	23	24	25	26	27	28	29
	30	31	32	33	34	35	36	37	38	39
	40	41	42	43	44	45	46	47	48	49
	50	51	52	53	54	55	56	57	58	59
	60	61	62	63	64	65	66	67	68	69
	70	71	72	73	74	75	76	77	78	79
	80	81	82	83	84	85	86	87	88	89
	90	91	92	93	94	95	96	97	98	99

Figure 3.4.13-3: Digital Pattern of 2.

Starting with 2 and adding 2 sequentially, you get every other even number following 2.

Note: this results in continuous vertical values of 0, 2, 4, 6, and 8.

3.4.13.4: Multiplication Pattern for 3.

First, make a copy of the multiplication table, and cut the columns into strips. Layout out each column on a table or tape them to a whiteboard.

3	0	1	2	**3**	4	5	**6**	7	8	**9**
	10	11	**12**	13	14	**15**	16	17	**18**	19
	20	**21**	22	23	**24**	25	26	**27**	28	29
	30	31	32	**33**	34	35	**36**	37	38	**39**
	40	41	**42**	43	44	**45**	46	47	**48**	49
	50	**51**	52	53	**54**	55	56	**57**	58	59
	60	61	62	**63**	64	65	**66**	67	68	**69**
	70	71	**72**	73	74	**75**	76	77	**78**	79
	80	**81**	82	83	**84**	85	86	**87**	88	89
	90	91	92	**93**	94	95	**96**	97	98	99

Figure 3.4.13-4: Digital Pattern of 3.

Starting with 3 and adding 3 sequentially, you get the third number following 3. This is the multiplication column of 3 as in Section 3.4.1.6 and the row-by-row sequence above.

Note: this results in a continuous diagonal set of blue values.

3.4.13.5: Multiplication Pattern for 4.

First, make a copy of the multiplication table and cut the columns into strips. Layout each column on a table or tape them to a whiteboard.

4	0	1	2	3	4	5	6	7	8	9
	10	11	12	13	14	15	16	17	18	19
	20	21	22	23	24	25	26	27	28	29
	30	31	32	33	34	35	36	37	38	39
	40	41	42	43	44	45	46	47	48	49
	50	51	52	53	54	55	56	57	58	59
	60	61	62	63	64	65	66	67	68	69
	70	71	72	73	74	75	76	77	78	79
	80	81	82	83	84	85	86	87	88	89
	90	91	92	93	94	95	96	97	98	99

Figure 3.4.13-5: Digital Pattern of 4.

Starting with 4 and adding 4 sequentially, you get the fourth number following 4 or every other Digital Pattern of 2 in Figure 3.4.13-3. This is the multiplication column of 4 as in Section 3.4.1.5 and the row-by-row sequence above.

Note: this results in off-set bottom-left to upper-right diagonal and off-set vertical values of 4 and 8.

3.4.13.6: Multiplication Pattern for 5.

First, make a copy of the multiplication table, and cut the columns into strips. Layout out each column on a table or tape them to a whiteboard.

5	0	1	2	3	4	5	6	7	8	9
	10	11	12	13	14	15	16	17	18	19
	20	21	22	23	24	25	26	27	28	29
	30	31	32	33	34	35	36	37	38	39
	40	41	42	43	44	45	46	47	48	49
	50	51	52	53	54	55	56	57	58	59
	60	61	62	63	64	65	66	67	68	69
	70	71	72	73	74	75	76	77	78	79
	80	81	82	83	84	85	86	87	88	89
	90	91	92	93	94	95	96	97	98	99

Figure 3.4.13-6: Digital Pattern of 5.

Starting with 5 and adding 5 sequentially, you get the fifth number following 5. This is the multiplication column of 5 as in Section 3.4.1.5 and the row-by-row sequence of the multiplication of 5 above.

Note: this results in continuous vertical values of 5 is not an even value, but its double, 10, is even.

3.4.13.7: Multiplication Pattern for 6.

First, make a copy of the multiplication table, and cut the columns into strips. Layout each column on a table or tape them to a whiteboard.

6	0	1	2	3	4	5	**6**	7	8	9
	10	11	**12**	13	14	15	16	17	**18**	19
	20	21	22	23	**24**	25	26	27	28	29
	30	31	32	33	34	35	**36**	37	38	39
	40	41	**42**	43	44	45	46	47	**48**	49
	50	51	52	53	**54**	55	56	57	58	59
	60	61	62	63	64	65	**66**	67	68	69
	70	71	**72**	73	74	75	76	77	**78**	79
	80	81	82	83	**84**	85	86	87	88	89
	90	91	92	93	94	95	**96**	97	98	99

Figure 3.4.13-7: Digital Pattern of 6.

Starting with 6 and adding 6 sequentially, you get the sixth number following 6. This is the multiplication column of 6 as in Section 3.4.1.6 and the row-by-row sequence of the multiplication of 6 above.

3.4.13.8: Multiplication Pattern for 7.

Make a copy of the multiplication table, and cut the columns into strips. Layout each column on a table or tape them to a whiteboard.

7

0	1	2	3	4	5	6	**7**	8	9
10	11	12	13	**14**	15	16	17	18	19
20	**21**	22	23	24	25	26	27	**28**	29
30	31	32	33	34	**35**	36	37	38	39
40	41	**42**	43	44	45	46	47	48	**49**
50	51	52	53	54	55	**56**	57	58	59
60	61	62	**63**	64	65	66	67	68	69
70	71	72	73	74	75	76	**77**	78	79
80	81	82	83	**84**	85	86	87	88	89
90	**91**	92	93	94	95	96	97	**98**	99

Figure 3.4.13-8: Digital Pattern of 7.

Starting with 7 and adding 7 sequentially, you get the seventh number following 7. This is the multiplication column of 7 as in Section 3.4.1.7 and the row-by-row sequence of the multiplication of 7 above.

3.4.13.9: Multiplication Pattern for 8.

Make a copy of the multiplication table, and cut the columns into strips. Layout out each column on a table or tape them to a whiteboard.

8

0	1	2	3	4	5	6	7	**8**	9
10	11	12	13	14	15	**16**	17	18	19
20	21	22	23	**24**	25	26	27	28	29
30	31	**32**	33	34	35	36	37	38	39
40	41	42	43	44	45	46	47	**48**	49
50	51	52	53	54	55	**56**	57	58	59
60	61	62	63	**64**	65	66	67	68	69
70	71	**72**	73	74	75	76	77	78	79
80	81	82	83	84	85	86	87	**88**	89
90	91	92	93	94	95	**96**	97	98	99

Figure 3.4.13-9: Digital Pattern of 8.

Starting with 8 and adding 8 sequentially, you get the eight number following 8. This is the multiplication column of 8 as in Section 3.4.1.8 and the row-by-row sequence of the multiplication of 8 above.

3.4.13.10: Multiplication Pattern for 9.

First, make a copy of the multiplication table, and cut the columns into strips. Layout out each column on a table or tape them to a whiteboard.

9										
	0	1	2	3	4	5	6	7	8	9
	10	11	12	13	14	15	16	17	18	19
	20	21	22	23	24	25	26	27	28	29
	30	31	32	33	34	35	36	37	38	39
	40	41	42	43	44	45	46	47	48	49
	50	51	52	53	54	55	56	57	58	59
	60	61	62	63	64	65	66	67	68	69
	70	71	72	73	74	75	76	77	78	79
	80	81	82	83	84	85	86	87	88	89
	90	91	92	93	94	95	96	97	98	99

Figure 3.4.13-10: Digital Pattern of 9.

Starting with 9 and adding 9 sequentially, you get the nine number following 9. This is the multiplication column of 9 as in Section 3.4.1.9 and the row-by-row sequence of the multiplication of 9 above. Why the diagonal, since the 10x10 matrix is 0 to 9 in each row, and 9 is one less than 10, the solution will offset by 1 in each subsequent row down.

Note: this results in diagonal values of 9 value and its multiplication values diagonally.

3.4.13.11: Summary.

Each value 0 to 9 has a unique geometric pattern or spacing. The 2 is in a vertical or column pattern, equally alternating by 2s from 0 to 9. The 3 and 4 is a diagonal, moving up to the right. The 5 is in a vertical or column pattern, equally spaced by 5s from 0 to 9. The 6, 7, 8 is a diagonal, moving up to the right. And that leaves the 9 value as a single diagonal, moving up to the right corner.

3.4.14: Pattern 14 – Digit Sum.

This is a unique approach to multiplication that isn't often taught. However, it's a pattern that verifies the results of any multiplication calculation and its product.

Digit Sum Rule: *The product of the digit-sum of a multiplicand and multiplier is the same as the digit sum of their product.*

3.4.14.1 Digit Sum Examples.

In Figure 3.4.14-1 is the multiplication of 32 x11. The multiplicand being 32, the multiplier 11, and the product being 352.

$$\left.\begin{array}{r} 32 \Rightarrow 3+2 \Rightarrow 5 \\ \underline{\times 11} \Rightarrow 1+1 \Rightarrow 2 \end{array}\right\} \Rightarrow 5\times2 = 10 = 1+0 \text{ or } \textcircled{1}$$
$$352 \Rightarrow 3+5+2 = 10 = 1+0 \text{ or } \textcircled{1}$$

Figure 3.4.14-1: 32 x11.

The multiplicand digit sum is 3 + 2 = 5 and the multiplier digit sum is 1 + 1 = 2. Since these are multiplied, it is intuitive that the digit sums should be multiplied: 5 x 2 = 10 or 1 + 0 = 1. The product digit sum is 3 + 5 + 2 = 10 or 1 + 0 = 1. Since the product digit sum is 1 and the product of the multiplicand and multiplier is 1, this is an excellent check on your multiplication work.

In Figure 3.4.14-2 is the multiplication of 80 x13. The multiplicand being 80, the multiplier 13, and the product being 1,040.

$$\left.\begin{array}{r} 80 \Rightarrow 8+0 \Rightarrow 8 \\ \underline{\times 13} \Rightarrow 1+3 \Rightarrow 4 \end{array}\right\} 8\times4 = 32 \text{ or } 3+2 = \textcircled{5}$$
$$1040 \Rightarrow 1+0+4+0 = \textcircled{5}$$

Figure 3.4.14-2: 80 x13.

The multiplicand digit sum is 8+0 = 8 and the multiplier digit sum is 1+3 = 4. Since these are multiplied, it is intuitive that the digit sums should be multiplied: 8x4 = 32 or 3+2 = 5. The product digit sum is 1+0+4+0 = 5. Since the product digit sum is 5 and the product of the multiplicand and multiplier is 5, a check on the multiplication is in agreement. In Figure 3.4.14-3 is the multiplication of 593 x106. The multiplicand is 593, the multiplier 103, and the product is 62,858.

```
  593   => 5+9+3 = 17  or 8 ]
x 106   => 1+0+6 =  7       ] 8x7 = 56 = 5+6 = 11 or (2)
62,858 => 6+2+8+5+8 = 29 or 2+9=11 or (2)
```

Figure 3.4.14-3: 593 x106.

The multiplicand digit sum is 5+9+3 = 17 or 1+7 = 8 and the multiplier digit sum being 1+0+6 = 7. Since these are multiplied, it is intuitive that the digit sums should be multiplied: 8x7 = 56 or 5+6 = 11 or 1+1 = 2. The product digit sum is 6+2+8+5+8 =29 or 2+9 = 11 or 1+1 = 2. Since the product digit sum is 2 and the product of the multiplicand and multiplier is 2, this shows a check on the multiplication.

In Figure 3.4.14-4. is the multiplication of 213 x90. The multiplicand is 213, the multiplier 90, and the product is 23,370.

```
   213  => 2+1+3 =>  6 ] 6*9 = 54 or 5+4 = (9)
  x 90  =>     9+0 => 9 ]
26,370  => 2+6+3+7+0  = 18 or 1+8 = (9)
```

Figure 3.4.14-4: 213 x90.

The multiplicand digit sum is 2+1+3 = 6 and the multiplier digit sum is 9+0 = 9. Since these are multiplied, it is intuitive that the digit sums should be multiplied: 6x9 = 54 or 5+4 = 9. The product digit sum is 2+6+3+7+0 =18 or 1+8 = 9. Since the product digit sum is 9 and the product of the multiplicand and multiplier is 9, the product is correct, and the check of the matching digit sum is correct.

In Figure 3.4.14-5 is the multiplication of 877 x 27. The multiplicand is 877, the multiplier is 27, and the product is 23,679.

```
   877  => 8+7+7 = 2+2 = 4 ] 4*9 = 36 or 3+6 = (9)
  x 27  =>      2+7 = 9    ]
23,679  => 2+3+6+7+9 = 27 =>2+7 = (9)
```

Figure 3.4.14-5: 877 x27.

The multiplicand digit sum is 8+7+7 = 22, which reduces to 2+2 or 4. The multiplier digit sum is 2+7 = 9. Since these are multiplied, it is intuitive that the digit sums should be multiplied: 4x9 = 36 or 3+6 = 9.

The product digit sum is 2+3+6+7+9 =27 or 2+7 = 9. Since the product digit sum is 9 and the product of the multiplicand and multiplier is 9, the product is correct. This is an excellent check on your multiplication work.

3.4.14.2 Digit Sum Summary (*simplified*)

As stated in the Digit Sum Rule is true, but there is a shortcut that is also a pattern of the rule. It can be seen that any digit sum that is 9 is also zero. So any digit of 9, or digits that add up to 9, can be ignored.

For example:

In Figure 3.4.14-1 above: The digit sum never includes a digit sum of 9 or numbers that add up to 9. So the shortcut does not apply.

In Figure 3.4.14-2 above: The digit sum never includes a digit sum of 9 or numbers that add up to 9. So the shortcut does not apply.

In Figure 3.4.14-3 above: The digit sum of the the multiplicand is 5+9+3 = 17 or 8, but drop the 9 and 5+3 = 8. Same answer.
The multiplier is 1+0+6 =7 and never adds up to 9. The product of these is 8x7 which is also 2. But, the product digit sums to 29, the 9 can be dropped, which leaves 2. Both solutions are 2 and that is a check.

In Figure 3.4.14-4 above: The digit sum of multiplicand is 2+1+3 = 6 and the multiplier is 9+0 =9 or zero. The product is 6x0 or zero. And the product digit sum is made up of: (3+6)+(2+7)+(9) are all sums of 9, so 0. Both solutions are zero and a check.

In Figure 3.4.14-5 above: The product of the multiplicand and multiplier is 36, which adds to 9, so it is zero. And the product digit sum is made up of: (3+6)+(2+7)+(9) are all sums of 9, so zero. So, both are zero, they match and there was less work to get there.

3.4.15: Pattern 15 – Digital Root (DR), Odd and Even Patterns.

In this pattern, the approach is to organize the individual columns of the multiplication table from the easiest to more difficult patterns of multiplication columns. This is a unique approach for understanding place value digits, 0 to 9, and their uses.

The digital root (DR) is the place value sum of the digit values. For example, 372 is 3 hundreds + 7 tens + 2 ones, so the DR is 3 + 7 + 2 = 12, and then 1 + 2 = 3. For any single-digit (1s place value) number, its DR is just that number. For any multi-digit number greater than 9, the DR is the sum of its place value digits.

64 => 6+4 = 10 = 1+0 => 1

16 => 1+6 = 7 => 7

507 => 5+0+7 = 12 => 1+2 => 3

198 =? 1+9+8 = 18 => 1+8 => 9

8915 => 8+9+1+4 = 22 => 2+2 => 4

1275 => 1+2+7+5 =15 => 6

This sets the stage for the next sections of Pattern 12 and how a DR can be used for patterns in the sticks or in addition and multiplication.

3.4.15.1: Digital Root for 0, 1, 5.

So, to start out we look at the natural numbers of: 0, 1, 2, 3, 4, 5, 6, 7, 8, 9.

Now in Section 3.4.1.2, we learned that any number (0 to 9) times zero (0) is just that zero. This is called the multiplication "Properties of Zero", and can be easily seen in the Tally-x stick of zero, Figure 3.4.1-2.

We also saw in Section 3.4.1.3, the multiplication "Properties of One", where any *number* (0 to 9)x1 is that *number* (0 to 9). This is called the multiplication "Proper of One", and can be easily seen in the Tally-x stick of zero, Figure 3.4.1-2.

So, that leaves us the natural numbers of: ~~0~~, ~~1~~, 2, 3, 4, 5, 6, 7, 8, 9.

We also saw in Section 3.4.1.4, the multiplication pattern of 5, where any *number* (0 to 9)x5 has a 1s place value of 0 or 5, see Figure 3.4.1-4. It is a good time to re-read that section.

So, that leaves us with the natural numbers of: ~~0~~, ~~1~~, 2, 3, 4, ~~5~~, 6, 7, 8, 9.

In summary, we now can see that the DRs have properties as well:

- … must be between 0 and 9.
- … that are 5 or 0 can be divided by 5.
- … that are 0 can be divided by 10.

So just as in earlier sections, 3.4.1.5; 3.4.1.6; and 3.4.1.7 we need to explore those DRs.

3.4.15.2: Digital Root for 2, 4, 8

~~0~~, ~~1~~, 2, 3, 4, ~~5~~, 6, 7, 8, 9.

Let's look at the natural numbers 2, 4, 8. As seen in Section 3.4.1.5, these numbers are related by the double and triple of 2. These numbers are all even and as such are also divisible by 2, whereas 4 and 8 can be divided by 4. This is the definition of an even number.

We can now add another DR property to our list:

- DR that is 2, 4, or 8 is even and can be divided by 2.

3.4.15.3: Digital Root for 3, 6, 9

~~0~~, ~~1~~, ~~2~~, 3, 4, ~~5~~, 6, ~~7~~, ~~8~~, 9.

In section 3.4.1.6 we see the 3 doubled (6) and tripled (9) values. Therefore, any number ending with a DR of 3, 6, or 9 is divisible by 3.

We can now add another DR property to our list:

- DR that is 3, 6, or 9 is odd and can be divided by 3.

3.4.15.4: Digital Root for 7.

~~0~~, ~~1~~, ~~2~~, ~~3~~, 4, ~~5~~, 6, 7, ~~8~~, ~~9~~.

In section 3.4.1.7 we see the 7 has an issue that "… only the seven needs a new approach …". This will not be covered not by DRs, but in Section 13.

3.4.15.5: Digital Root - addition application.

In the above sections, we see the application of the Digital Root (DR) in single-digit multiplication to identify division by whole values. But, it can also be used to check the addition of rows of values.

In Figure 3.4.15-1, we address the sum of two numbers, both being 2-digits: 32 and 11. Of course, the answer is 43, but we will use the DRs of each row to validate that this is correct:

```
  32  => 3+2 =>  5 ]
+ 11  => 1+1  => 2 ]=> 5+2 => (7)
  43  => 4+3  => (7)
```

Figure 3.4.15-1: 32+11.

In Figure 3.4.15-2, we address the sum of two numbers, both being 2-digits: 80, and 13.

```
  80 => 8+0 =>  8 ]
+ 13 => 1+3  => 4 ]= 8+4 = 12 => 1+2 => (3)
  93 => 9+3  => 12 => 1+2 => (3)
```

Figure 3.4.15-2: 80+13.

In Figure 3.4.15-3, we address the sum of three numbers, both being 3 digits: 106, 593, and 712.

```
  106  => 1+0+7 =>  7 ]
  593  => 5+9+3 => 17 ]= 7+(1+7)+(1+0) = 16 => (7)
+ 712  => 7+1+2  => 10 ]
 1411  => 1+4+0+1  => (7)
```

Figure 3.4.15-3: 106+593+712.

In Figure 3.4.15-4, we address the sum of two numbers, being a mix of 2 and 3digits: 213, 567, and 90.

```
  213  => 2+1+3 =>  6 ]
  567  => 5+6+7 => 18 ]= 6+(1+8)+(9+0) = 2+4 => (6)
+  90  =>     9+0  => 9 ]
  870  => 8+7+0  = 15 => (6)
```

Figure 3.4.15-4: 213+567+90.

In Figure 3.4.15-5, we address the sum of two numbers, both being 2-digits: 258, 27, and 877.

```
  258  => 2+1+3 => 6 ]
   27  =>    2+7 => 9 ]= 6+9+(2+2) = 1+9 =>10 => (1)
+ 877  => 8+7+7 =>22 ]
 1162  => 1+1+6+2 = 10 =>1+0 => (1)
```

Figure 3.4.15-5: 258+27+877.

3.4.15.6: Summary

We have addressed the sum of two numbers, both being 2 digits: 32 and 11.

We now have all the DR Properties:

1. … must be between 0 and 9.
2. … that are 5 or 0 can be divided by 5.
3. … that are 0 can be divided by 10.
4. … that is 2, 4, or 8 is even and can be divided by 2.
5. … that is 3, 5, or 9 is odd and can be divided by 3.

3.4.16: **Pattern 16** – Tesla's Digital Root Circle.

This is a special pattern developed by Nicholas Tesla and has applications not only in the multiplication tables but in theoretical and applied sciences. The 1, 2, 4, 8, 7, and 5 patterns are based on Pattern 15 of the Digital Root. In this 3, 6, 9 geometric pattern, which is similar to the numeric Pattern 3.4.1.6 of the multiplication columns of 3, 6, 9. All this will be covered in the next two sub-sections.

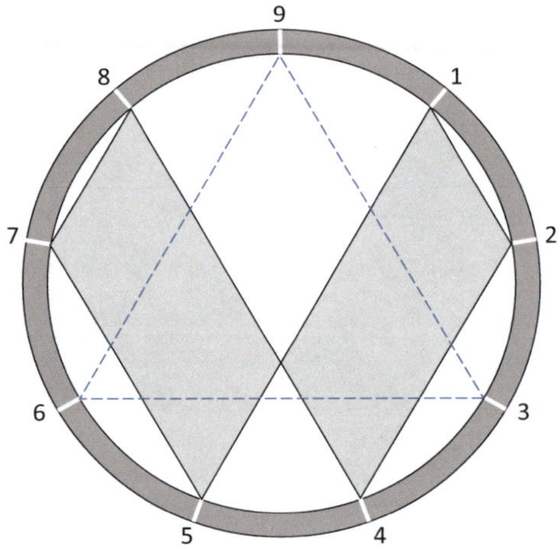

Figure 3.4.16-1: Tesla's Circle.

3.4.16.1: The 1s Doubled Pattern or the 1,2,4,8,7,5 Digital Root Pattern.

In Figure 3.4.16-1, you see the gray "W" shape which is a linear progression of 1 and its doubling sequence of 1, 2, 4, 8, 16, 32, and back to 1 in the above Figure. But, the DR values read as: 1, 2, 4, 8, 7, 5, and back to 1. This is also true if you *half* the sequential values, starting with 1, where the solution is 1, 0.5, 0.25, 0.125, 0.0625, and 0.03125, where the DR for each is 1, 5, 7, 8, 4, and 2, which is the reverse of the "W" shape using doubling. In both cases, the *doubling* or *halving* sequence could continue, but the DR pattern would only repeat what is in the "W" pattern circle.

What is significant in this DR pattern is still to be discovered, but one observation is that it contains two interesting features found in 1) Section 3.4.1.3 One (1), Section 3.4.1.7: Seven (7), and Section 3.4.1.4 Five (5) on single value sequences and 2) that of the multiple sequence in Section 3.4.1.5: 2-4-6-8. At this time, why these are related *other than their linear progression is still to be discovered according to Tesla.*

3.4.16.2: The 3-6-9 Digital Root Pattern.

In Figure 3.4.16-1, you see the blue equilateral triangle shape for the values of 3, 6, 9. In this case, the pattern is non-linear in that the spacing is growing with the *doubling* and *tripling* of the initial value of 3. We discussed in in Section 3.4.1.6: 3-6-9.

Tesla and some current scientists believe that these three numbers (3-6-9) are related to some higher dimension and therefore new math format we have yet to learn. Hey, it wasn't 2500 years since the ICV understood Geometry, that the Greeks re-invented it. And Calculus was not discovered by England and Europe 1600s AD, or 4600 years after understood by the ICV.

3.4.16.3: 3-6-9 pattern put into action.

> "*If you know the Magnificence of the 3-6-9, then you would have a key to the Universe.*" Nichola Tesla Quote.

Tesla was fascinated with the 3 6 9 number because he found it in so many math and other sciences. For example:

Mathematics: The number 3 is, see *reference 3.6*:

- … is the first prime or "father prime" number,
- … is the only number that equals the sum of all preceding numbers (0 + 1 + 2 = 3),
- … when squared, completes the single-digit sequence of 9.

"Omne trium perfectum"
'*everything that comes in 3s is perfect.*'

Geometry: the circle has 360 degrees to it and the Digital Root (DR) of that is 9. But,

- divide a circle in half and you get 180 degrees or a DR of 9,
- divide the half circle in half and you get 90 degrees or a DR of 9,
- Continue dividing each smaller circle and you get 9.
  - 45° ⇨ DR = 4+5 ⇨ 9,
  - 22.5° ⇨ DR = 2+2+5 ⇨ 9,
  - 11.25° ⇨ DR = 1+1+2+5 ⇨ 9,
  - 5.625° ⇨ DR = 5+6+2+5 ⇨ 9,
  - 2.81254° ⇨ DR = 2+8+1+2+5+4 ⇨ 9,
  - 1.40625° ⇨ DR = 1+4+0+6+2+5 ⇨ 9.

This sequence continues on, but also applies to triangles (3-sides) and rectangles (4-sides), pentagon (5-sides), Hexagon (6-sides), Heptagon (7-sides), Octagon (8-sides), etc.

Writing: There is a writing principle, see reference 3.7, that suggests that three entities used in events are more humorous, satisfying, or effective than any other number. The audience is also more likely to remember the information because these combine both brevity and rhythm with having the smallest amount of information to create a pattern. This can be used in slogans, film titles, advertising, stories/movies, and abbreviations. For example:

- … 'Life, Liberty and the Pursuit of Happiness' – U.S. Declaration of Independence,
- … 'The Lion King'; 'The Dark Knight'; 'The Ten Commandments', …
- … 'Just do it'(Nike); 'Quicker Picker Upper' (Bounty) ; 'I'm Lovin' it' (McDonalds), …
- … 'Three Little Pigs'; 'The Shoes Books'; 'Winnie-the-Pooh'; 'The Little Prince'.

Sciences:

Electricity: Nickola Tesla's Alternating Current (AC) in the U.S.A. won over the Direct Current of Thomas Edison (DC) could not be transmitted over long distances.

voltage (US):	120 volts	DR => 1+2+0	=9
frequency (US):	60 hertz	DR => 6+0	=9

See Reference 3.8.

Physics:

light speed:	186282 miles/sec	DR => 1+8+6+2+8+2	=9

Astronomy:

Earth Diameter:	7,920 miles	DR => 7+9+2+0	=9
Sun Diameter:	864,000 miles	DR => 8+6+4+0+0+0	=9
Moon Diameter:	2,160 miles	DR => 2+1+6+0	=9

Time:

Seconds per Hour:	3,600	DR => 3+6+0+0	= 9
Seconds per Day:	86,400	DR => 8+6+4+0+0	= 9
Seconds per Week:	604,800	DR => 6+0+4+8+0+0	= 9
Seconds per Month:	2,592,000	DR => 2+5+9+2+0+0	= 9
Seconds per Year:	31,536,000	DR => 3+1+5+3+6+0+0+0	= 9
Minutes per Day:	1,440	DR => 1+4+4+0	= 9
Minutes per Week:	10,080	DR => 1+0+0+8+0	= 9
Minutes per Month:	43,200	DR => 4+3+2+0	= 9
Minutes per Year:	525,600	DR => 5+2+5+6+0+0	= 9

Religion:

Even religion pays respect in different scriptures using 3-6-9 in many ways. But, I will leave that to you and your parents to discuss and discover in your readings.

3.4.17: **Pattern 17**–Odd and Even Digital Root Patterns.

In this pattern, the approach is to organize the individual columns of the multiplication table from the easiest to more difficult patterns of multiplication columns. This is not based on prior memorization or Pattern 1. Because of the multiplication Rule of Zero (0) we can leave zeo out of pattern 17, which leaves the three sections below of: 2, 4, 8 / 3, 6, 9 / 1, 5, 7. And we will focus on the even and odd values to find a new pattern.

3.4.17.1: Digital Root (DR) for 2, 4, 8.

In this example, the values of 2, 4, and 8 have a cross relationship where 4 is the double and 8 is the triple of the first value 2, as you may remember in Pattern 1. See Section 3.4.1.5.

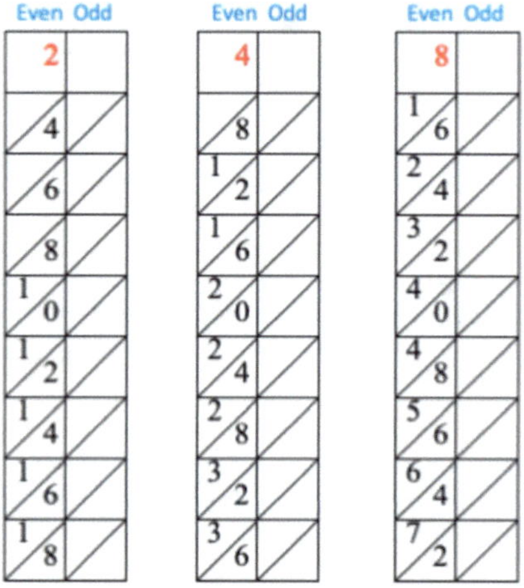

Figure 3.4.17-1: Digital Root of 2, 4, 8

Looking at Figure 3.4.17-1 you can see that the values are all even, so all values go in the left or even column. This makes sence as any odd or even (1, 2, 3, 4, 5, 6, 7, 8, 9) times an even number (2, 4, 8) is an even number. See Properties of Multiplication for even and odd numbers.

Now let's look at the DR of 2, 4, 8:

- DR (2): 0+2, 0+4, 0+6, 0+8, 1+0, 1+2, 1+4, 1+6, 1+8=> 2, 4, 6, 8, 1, 3, 5, 7, 9.
- DR (4): 0+4, 0+8, 1+2, 1+6, 2+0, 2+4, 2+8, 3+2, 3+6=> 4, 8, 3, 7, 2, 6, 1, 5, 9.
- DR (8): 0+2, 0+4, 0+6, 0+8, 1+0, 1+2, 1+4, 1+6, 1+8=> 8, 7, 6, 5, 4, 3, 2, 1, 9.

Now the pattern of the DR is interesting.

- For 2: starts with 2 and ends with 9, starts with all even values, and then all odd values,
- For 4: starts with 4 and ends with 9, alternates with 2 even values (4, 8), 2 odd values (3, 7), then 2 even values (2, 6), 2 odd values (1, 5), and ends with 9.
- For 8: starts with 8 and ends with 9, starts 8 and then counts down (8, 7, 6, 5, 4, 3, 2, 1, 9).

So, the pattern is that is always starts with the root value of multiplication, but the DR ends in 9.

3.4.17.2: Digital Root for 3, 6, 9.

In this example, the values of 3, 6, 9 have a cross relationship where 6 is the double and 9 is the triple of the first value, 3. See Section 3.4.1.6.

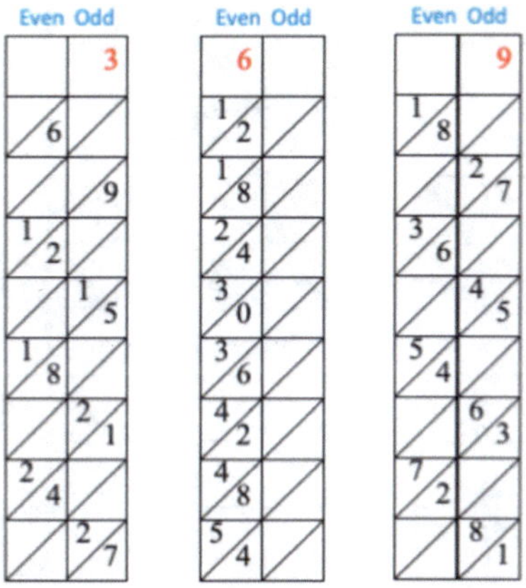

Figure 3.4.17-2: Digital Root of 3, 6, 9

Looking at Figure 3.4.17-2 you can see that the values are odd (3, 9) and one even (6). This is because the double of 3 is even (odd time even) and the triple of 3 is odd (odd times odd). See Properties of Multiplication for even and odd numbers. You will also see that 3, and 9 are alternating odd and even, between root values are odd and the multiplier is odd. Whereas when the root value is multiplied by an even value you get an even value. The same Properties of Multiplication for even and odd numbers.

Now let's look at the DR of 3, 6, 9:

- DR (3): 0+3, 0+6, 0+9, 1+2, 1+5, 1+8, 2+1, 2+4, 2+7 => 3, 6, 9, 3, 6, 9, 3, 6, 9.
- DR (6): 0+6, 1+2, 1+8, 2+4, 30+, 3+6, 4+2, 4+8, 5+4 => 6, 3, 9, 6, 3, 9, 6, 3, 9.
- DR (9): 0+9. 1+8. 2+7. 3+6. 4+5. 5+4. 6+3. 8+2. 8+1 => 9, 9, 9, 9, 9, 9, 9, 9, 9.

Now the pattern of the DR is similar to 2, 4,8 in the first pattern and matches that of Pattern 1 in Section 3.4.1.6. This makes you think:

- For 3: starts with 3 and ends with 9, is the DR of the first three values, repeated three times.
- For 6: starts with 6 and ends with 9, is the DR of the first three values, repeated three times.
- For 9: starts with 9 and ends with 9, is the DR of the first value, repeated nine times.

So, the pattern is that it always starts with the root value of multiplication, but the DR ends in 9. And, the first of three values are the DR values of the remaining multiplications.

3.4.17.3: Digital Root for 1, 5, 7.

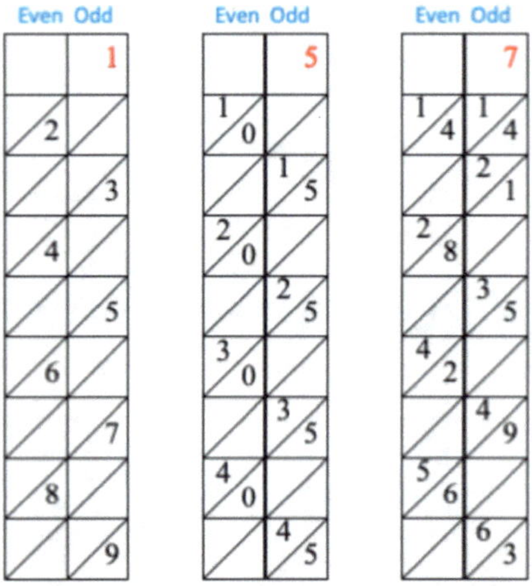

Figure 3.4.17-3: Digital Root of 1, 5, 7

In this section, there is no relationship between 1, 5, 7. The pattern is internal to each value.

Digital Root of 1.

- starts with 1 and ends with 9, i.e., 1, 2, 3, 4, 5, 6, 7, 8, 9

This is seen as the multiplier values and is explained in Section 3.4.1.3 as the Properties of 1.

Digital Root of 5.

- starts with 5 and ends with 9, i.e., 5, 1, 6, 2, 7, 3, 8, 4, 9

The sequence stars with root value (5) and alternates with the odd multiplier by counting up: 5, 6, 7, 8, 9. Whereas the second value starts with the root value's multiplier (1) and counts up: 1, 2, 3, 4.

Digital Root of 7.

- starts with 7 and ends with 9, i.e., 7, 5, 3, 1, 8, 6, 4, 2, 9

The sequence starts with root value (7) and then counts down by 2s: 7, 5, 3, 1. Adding 7 to 1, you get 8 and counts down: 8, 6, 4, 2 and ending with 9.

3.4.17.4: Summary.

The start and stop for each DR of 1, 5, and 7 will always start with the root value and end with 9. Just like all the other root values this is unique to Pattern 17.

3.5: Summary.

The author started with 8-patterns, and over the writing of this book, discovered three (3) more patterns, and the three Groupings. Can there be more – YES!

The author invites anyone to provide additional Column, Table, or Digit based patterns not shown in 2.3 for future Editions of this book.

Chapter 3 Endnotes.

3.1 Deleanu, Dumitru. (2011). The Dynamics of a Double Pendulum: Classic and Modern Approach, Annals of "Dunarea De Jos" University of Galati. Year III (XXXIV). 203-212.

3.2 Devaney, Robert Luke Devaney, "An Introduction to Chaotic Dynamical Systems," 2nd edition, 2003, ISBN 978-08133-4085-2

3.3 Odd: a number that is an integer that is not even or divisible by 2. (i.e., 1, 3, 5, 7, 9).

3.4 The square of a number is based on geometric origins. Where all other non-equal row and column values form a geometric rectangle, the equal row, and column values form a square.

3.5 A. Benjamin, M. Shermer, "Secrets of Mental Math", Three Rivers Press, 2006, ISBN 978-03073-3840-2

3.6 Kaleena Fraga, "Inside Nikola Tesla's 3, 6, 9 Obsession and the Unusual Theories It Spawned" 16 August 2021, https://allthatsinteresting.com/nikola-tesla-3-6-9 (last accessed 02 Dec 2022)

3.7 "Rule of Three (writing)", Wikipedia, https://en.wikipedia.org/wiki/Rule_of_three_(writing) (last accessed 02 Dec 2022)

3.8 Leonardo B Beretta, "Nicola Tesla and the 369 Code: Decoding God's Thumbprint", 2018, ISBN 978-19769-3279-3

Chapter 4: Multiplication

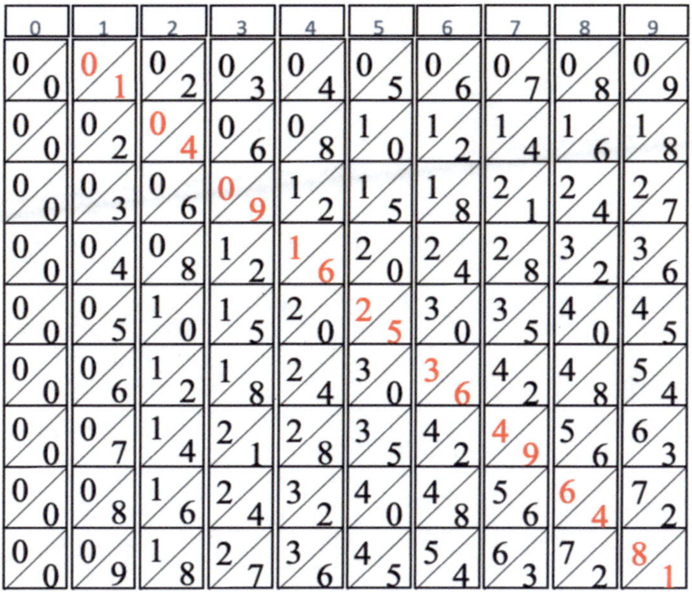

```
  7
x 3
 21
```

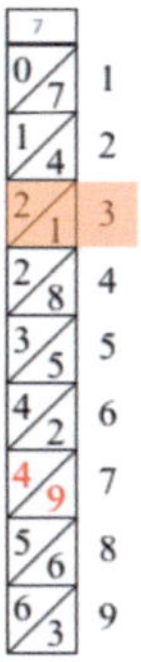

```
   97
  x 4
   28
  36
  388
```

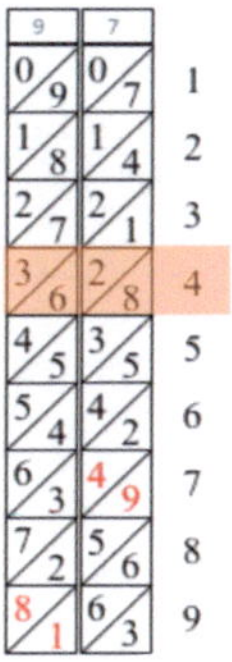

```
  9
x 7
 63
```

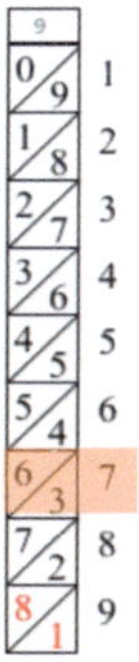

4.1: The Basics of Tally-x.

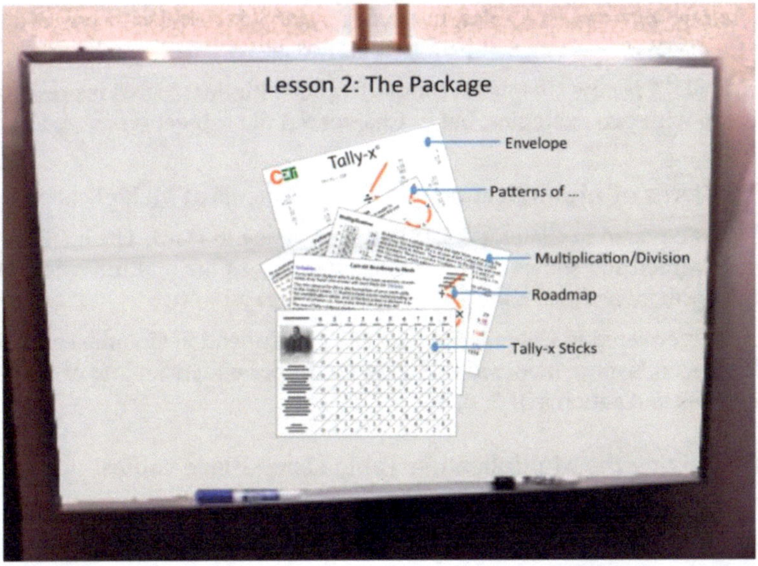

Figure 4.1-1: What came in the original Tally-x Package?

The original Tally-x package (ISBN 978-0-578-08779-5) is shown in Figure 4.1-1 The Tally-x sticks were first developed in the 1990s using Popsicle sticks, with the Tally-x stick card in 2009, the Tally-x package in 2011, and the slide presentation in 2013. This package came with four cards and an envelope, see below. This was complemented by a slide presentation that discussed how Tally-x works. Each lesson discussed a small part of what is in this book.

Card 1: Tally-x sticks ("Tally-x Sticks" card).

The Tally-x stick card was made of plastic and perforated to allow it to be snapped into ten columns, 1 to 9, and a stick that shows John Napier on one side and how to do a rudimentary single-digit multiplier (SDM) on the other side.

Card 2: Roadmap ("Roadmap" card).

The roadmap is covered in Chapter 1.4: Roadmap of Arithmetic and Chapter 1.5: Applying the Roadmap to the Curriculum of Math.

Card 3: Multiplication / Division ("Multiplication/Division" card).

Multiplication, SDM is now Chapter 4. Single-Digit Multiplier (SDM) is presented in the Lesson with two examples, but in Chapter 4.3 the subject is covered by many examples.

Division, SDM is now Chapter 5. Single-Digit Multiplier (SDM) is presented in the Lesson with two examples, but in Chapter 5.3 the subject is covered by many examples.

Card 4: Patterns of Multiplication ("Patterns of Multiplication" card).

Pattern I is covered in Chapter 4.3.1: Pattern 1- Easy to Hard. The book's version is more involved and shows a greater understanding of the multiplication patterns. This is why there are now a total of 11 patterns instead of two.

Pattern II is covered in Chapter 4.3.1.6: The Multiplication Columns of 9. The book's version is more involved and shows a deeper understanding of the relationships and patterns of 3, 6, 9.

4.1.1: Organizing the Multiplication Table Using Place values.

Each Tally-x stick is the multiplication table for the number at the top. For example, the stick with the number two (2) at the top is the multiplication table for 2, for multipliers 1 to 9. See Figure 4.1.1-1.

	0	1	2	3	4	5	6	7	8	9
1	0	1	2	3	4	5	6	7	8	9
2	0	2	4	6	8	10	12	14	16	18
3	0	3	6	9	12	15	18	21	24	27
4	0	4	8	12	16	20	24	28	32	36
5	0	5	10	15	20	25	30	35	40	45
6	0	6	12	18	24	30	36	42	48	54
7	0	7	14	21	28	35	42	49	56	63
8	0	8	16	24	32	40	48	56	64	72
9	0	9	18	27	36	45	54	63	72	81

Figure 4.1.1-1 – The Traditional Multiplication Column Table.

In Figure 4.1.1-2 is the traditional approach to introducing the multiplication table to students. Each column is presented in a single (one's value) digit or a double (ten's and one's value) digit. Therefore, the maximum place value of two 1-digit numbers results in a 2-digit number but is addressed later.

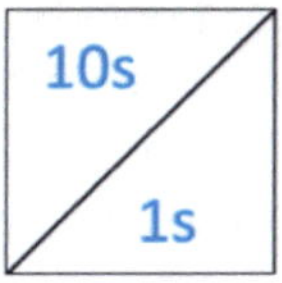

Figure 4.1.1-2 –Separating the 10s and 1s values, i.e., Lattice System.

The Lattice system, initially introduced by Leonardo da Pisa in the twelfth century, can be used alongside Tally-x sticks to obtain results, as shown in Figure 4.1.1-3. Compared to the traditional long-multiplication method described earlier, it functions similarly by carrying values into higher place columns of numbers. Specifically, it places the tens-place values of the ones column above the diagonal line in the lattice grid, creating the appearance that they would slide downward to the next column on the left. This mechanism ensures that high-place values are carried over correctly to the next column, akin to the principles of the Lattice system or through place-value shifting in long multiplication.

Leonardo documented the Lattice method based on teachings he received from Arab sources and earlier traditions. One drawback of the Lattice method is that users must memorize values to fill in the blank lattice grid. However, John Napier proposed a solution where lattice values could be pre-filled in columns, resulting in a straightforward and universal multiplication tool, depicted in Figure 4.1-4.

	0	1	2	3	4	5	6	7	8	9
1	0	1	2	3	4	5	6	7	8	9
2	/0	/2	/4	/6	/8	1/0	1/2	1/4	1/6	1/8
3	/0	/3	/6	/9	1/2	1/5	1/8	2/1	2/4	2/7
4	/0	/4	/8	1/2	1/6	2/0	2/4	2/8	3/2	3/6
5	/0	/5	1/0	1/5	2/0	2/5	3/0	3/5	4/0	4/5
6	/0	/6	1/2	1/8	2/4	3/0	3/6	4/2	4/8	5/4
7	/0	/7	1/4	2/1	2/8	3/5	4/2	4/9	5/6	6/3
8	/0	/8	1/6	2/4	3/2	4/0	4/8	5/6	6/4	7/2
9	/0	/9	1/8	2/7	3/6	4/5	5/4	6/3	7/2	8/1

Figure 4.1.1-3 – One Side of John Napier's Bones.

The difference between John Napier's Bone system and Tally-x sticks is that (1) the zero-row is removed, (2) the ones-row is the multiplicand, and (3) the is lattice is printed, as nine compliments, on both sides of the stick. This results in Figure 4.1.1-4.

However, John Napier's was printed on a bone with 4 sides in such a way as to maximize the number of repeat values. So, the Tally-x stick is maximized by using the 9's complement. The idea of this complement is that the multiplicand on one side plus the multiplicand on the opposite side must add up to nine (9). So, in Table 4.1.1-5, you see the multiplicand of side-1 is across the top row, and the multiplicand of side-2 is across the bottom row, all adding to 9.

0	1	2	3	4	5	6	7	8	9
9	8	7	6	5	4	3	2	1	0

Table 4.1.1-4 – The 9's Complement of the Tally-x top row.

Napier's Bones is a little more complicated but based on the same 9s complement on opposite sides. The difference is that Napier used 4 sides of ivory or 1-side of boxwood to carve his multiplication tables. Note in Figure 4.1.-5 that each side adds up vertically to 18, as both opposing sides A+C and B+D must add up to 9. However, there appears to be no pattern in the distribution of the numbers, where a uniform distribution would seem likely.

Sides	A	B	C	D	
1	0	1	9	8	18
2	0	2	9	7	18
3	0	3	9	6	18
4	0	4	9	5	18
5	1	2	8	7	18
6	1	3	8	6	18
7	2	4	7	5	18
8	2	3	7	6	18
9	2	4	7	5	18
10	3	4	6	5	18
	11	30	79	60	

A

Sides	A	B	C	D	
1	0	2	9	7	18
2	1	3	8	6	18
3	2	6	7	3	18
4	3	0	6	9	18
5	4	1	5	8	18
6	5	7	4	2	18
7	6	4	3	5	18
8	7	8	2	1	18
9	8	9	1	0	18
10	9	5	0	4	18
	45	45	45	45	

B

Figure 4.1.1-5 – John Napier used "A", but a more uniform pattern is "B".

This 9s complement is also found in the pattern in Chapter 2.3.3.6 The Multiplication of Columns of 3, 6, 9 where the 9-pattern always adds up horizontally to 9.

Napier's bones are the foundation of Tally-x as a tool. As stated, it has an advantage over the Lattice system in that "table memorization" is unnecessary, and it can be used for division. The sticks are a full block set of the Lattice multiplication series for each value. Think of it, as if, 123,456,789 (across the top) is multiplied by each value 0 to 9 (down the left side). Now you can see the transition of the Lattice method to John Napier's method.

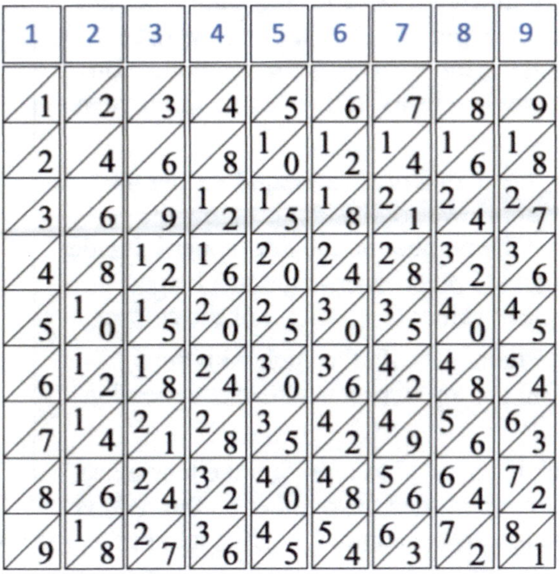

Figure 4.1.1-5 – The Tally-x Stick Columns.

In Figure 4.1.1-5 is a full set of sticks before they are separated into columns. This is why memorization is not necessary; it is a pattern, and its frequent use makes it intuitive to the user.

4.1.2: General Layout of the Tally-x sticks.

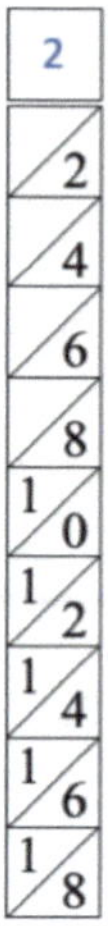

Figure 4.1.2-1
2's stick

Each of the Tally-x sticks, once separated from the set, represents its multiplication table for the number at the top. For instance, in Figure 4.1.2-1, the stick with the number two (2) at the top represents the multiplication table for 2. This can be observed by examining the rows on the stick.

In a single stick, the numbers are arranged so that the number on the right (lower right side) is below the diagonal line, and the number on the left (upper left side) is above the diagonal line. This arrangement helps distinguish the place values (1s, 10s, ...) of the column.

For example, in Figure 4.1.2-2, the stick with the number nine (9) at the top represents the multiplication table for 9.

If the single stick is intended to represent the tens place value of the top number, then the number on the right is the ones place value of that number, and the number on the left is the tens value.

Figure 4.1.2-2
9's stick

In each case, the multiplication table on the stick cannot exceed 100. For example, 9x9 equals 81, which is the highest possible number for a multiplication stick.

If the single stick is intended to represent a higher value, such as 100, then the number on the right is the 100's place value of the number, and the number on the left is the 1,000's place value. For example, the number 900 would be represented by the number 9 stick. Therefore, the previous example of 9x9 would be best represented by 900 x9, which is 81 plus the two zeros: 8100.

4.1.3: Use of the Tally-x sticks.

In the example in Figure 4.1.3-1 is the number 900, where the 9s stick is in the 100s place value, you can do that with your mind; however, you can use the 0s stick as a placeholder for the 10s and 1s value. Resulting in the 900 multiplication value.

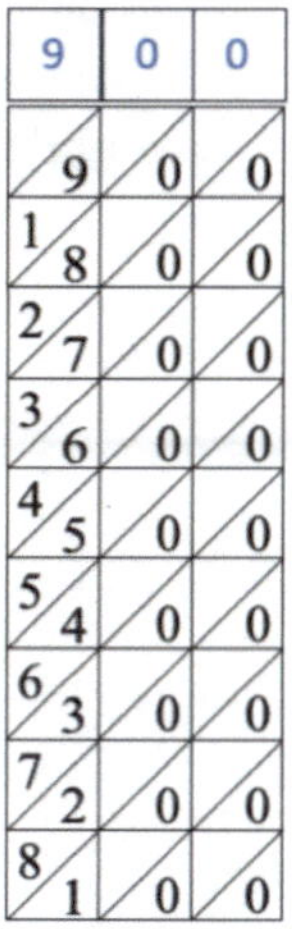

Figure 4.1.3-1
9 0 0 sticks

This principle of adding the 0 sticks to get the 900 number, quickly shows the method behind multiplication using the sticks. Add the next value to your column to create a large number at the top. This number at the top is called the multiplicand. The single-digit multiplier is simply the row number. For example, 900 x1 is 900.

There is no reason why people memorize the multiplication tables. However, I would venture to guess that most have never used multiplication tables in this manner. Once you have used the sticks in this manner and understand the pattern of the numbers, you will not have to memorize the 0 to 9 multiplication table, because you will understand this stick pattern. Like remembering a song.

Multiplication is simply the collection of the appropriate single-digit multiplication tables (i.e., sticks). You can create more significant numbers with more sticks. For example, in Figure 4.1.3-2, you have the multiplicand 29 which is a matter of laying down the 9-stick for the 1s place value and then to its left laying down the 2-stick for the 10s place value.

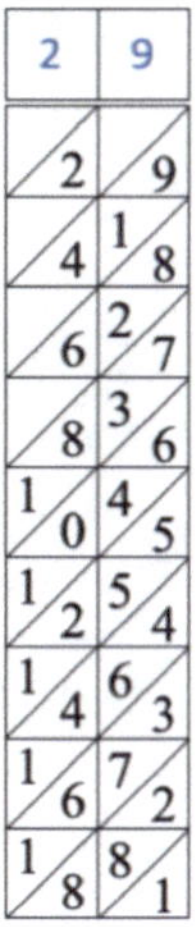

Figure 4.1.3-2
2 9 sticks

In the example to the left, the 2-stick, representing the 10s values, is to the right of the 9-stick, representing the 1s values. With the rows properly aligned left to right, the solution is a matter of finding the 6th row.

This could have been done with the multiplicand of 20 and then adding the 9; however, we saved work by combining the two sticks instead. So, the solution of 29 x6 is that the 1s value is 4, the 10s value is 2+5=7, and the 100s value is 1. Answer: 174. It is always best to create the answer from the 1s place value and progress to the highest place value, thereby allowing carries.

To handle two or more digit multipliers, is a matter of writing down the first multiplier (1s) answer, using the same multiplicand sticks, write down the next multiplier (10s) aligned under the first answer, however, shifted left by one zero. Continue for each multiplier value. Continue for each multiplier value, then add the place values for the answer.

4.2: Traditional Multiplication.

In both short and long multiplication are used for the vertically written calculations. The short multiplication method is used for one-digit multipliers only, and long multiplication is used for two- or more- digit multipliers. Both involve carry-over values to higher place values and use the same identifiers: Multiplicand, Multiplier, and Product. See Figure 4.2.1-1 for a quick overview.

4.2.1: Short Multiplication.

The basic identifiers of the values are:

- *Multiplicand*: a number that is multiplied by the multiplier, repeated value, upper number. I.e., 13 in Figure 4.2.1-1.
- *Multiplier*: number by which the multiplicand is multiplied, count the lower number. I.e., 3 in Figure 4.2.1-1.
- *Product*: the result of two numbers being multiplied, under the solution line. I.e., 39 in Figure 4.2.1-1.

Short multiplication is memory dependent on the Rote learning method discussed in Chapter 2.1.1. In Figure 4.2.1-1, the multiple numbers are arranged above the calculation bar, with the multiplicand over the multiplier. In general, it does not matter which number of the two multiplication values is the multiplicand or the multiplier.

```
 13  multiplicand
x 3  multiplier
 39  product
```

Figure 4.2.1-1
13 x3.

Example 1: 7 x8.

```
  7  multiplicand
x 8  multiplier
 56  product
```

Figure 4.2.1-2
7 x8.

Figure 4.2.1-2 is the process that the multiplier, eight (8), is first multiplied by the multiplicand, seven (7), in the 1s multiplicand place value. Then, using the Rote learning method to mentally calculate the resulting product. The answer would mentally be fifty-six (56).

Due to the Commutative Properties of multiplication, the same answer is universally obtained if the multiplicand and multiplier are switched. I.e., Where seven (7) is the multiplicand and the eight (8) is the multiplier. Therefore, either way, the product or answer is 56.

Example 2: 13 x2.

The process of Figure 4.2.1-3 is as simple, in that the two (2) are first multiplied by the three (3) in the 1s multiplicand place value. Then again using the Rote learning method, the answer would mentally be derived as six (6) which is entered as the 1s product place value.

13	multiplicand
x 2	multiplier
26	product

Figure 4.2.1-3
13 x2.

However, although the Commutative Properties of multiplication works here as well, it would have to be done in Long Multiplication because of the double-digit multiplier, 13. However, in either case, the product is still 26.

4.2.2: Long Multiplication.

Long multiplication using the same identifiers (i.e., Multiplicand, Multiplier, and Product) is also more memory dependent on the Rote learning method. In Figure 4.2.2-1, the multiple numbers are also arranged above the calculation bar, with the multiplicand over the multiplier. Here carry-overs are values in any single place value that is greater than 9. The difference is then added to the next higher place value.

Example 1: 27 x8.

The process of Figure 4.2.2-1 is that the eight (8) is first multiplied by the seven (7) in the 1s multiplicand place value. Then, using the Rote learning method, the answer would be fifty-six (56), which is entered below the multiplication bar. The process continues when the eight (8) is multiplied by the two (2) in the 10s multiplicand. Again, using the Rote learning method, the answer would be sixteen (16) in the 10s Product place value or shifted left by 1-place value. Note the space to the right.

27	multiplicand
x8	multiplier
56	
$1^{1}6$	
2 16	product

Figure 4.2.2-1
27 x8.

The product is then the addition of the two individual answers: six (6) for the 1s product value, five(5) plus six (6) or eleven (11) in the 10s product place valve, then ten (10) is subtracted from the eleven (11) value, and only the one (1) is recorded for the 10s product place value, and finally the one (1) is added to the carry-over to come to two (2) in the 100s product value. Therefore, the product is 216.

Example 2: 123 x26.

The process of Figure 4.2.2-2 is that the six (6) is first multiplied by the three (3) in the 1s multiplicand place value. Then the Rote learning method provides the mental answer of twenty-four (24), which is entered as the 1s place value product under the multiplier bar. Continuing with six (6) multiplied by the two (2) in the 10s multiplicand place value, resulting in twelve (12), which is written under the 24 value but 1-place value to the left (10s value). Then the six (6) is multiplied by the one (1) in the 100s multiplicand place value, resulting in the six (6) written under the 12 value. Now the two (2) is multiplied by the one (1) in the 100s multiplicand.

123	multiplicand
x26	multiplier
24	
$1^{1}2$	
6	
$2^{1}4\,6$	
3 2 04	product

Figure 4.2.2-2
123 x26.

Now for the 10s multiplier place value, the two (2) is multiplied by each of the multiplicand values. These will result in a single value each, or 246, which is placed below the 6 value, with no shifting, and a line is drawn beneath. The values are added up, with appropriate carry-overs for a final product of 3204.

4.3: Tally-x Stick Multiplication

4.3.1 Single-Digit Multiplier (SDM).

Multiplication is simply collecting the correct multiplicand Sticks and reading the correct row as the multiplier. For example, 26 x4 is a matter of collecting the 2-stick and the 6-stick and placing them side-by-side. The 2-stick in the tens place, and the 6-stick in the one's place. Then read down to the 4th row and read the numbers across. There is a 4 in the 1s column. There is an 8 and 2 in the two 10s columns, which add together for 10. Answer: 104.

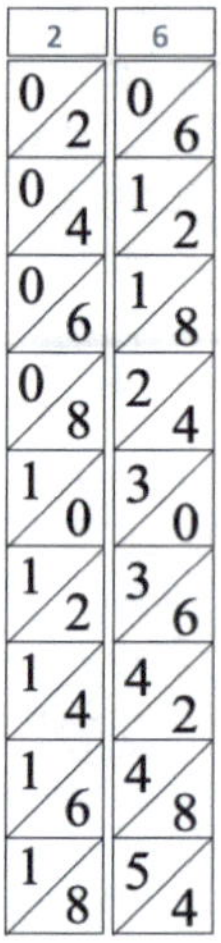

Figure 4.3.1-0
2 6 sticks.

This method works for any multiplicands, as a collection of sticks, no matter how many digits are in the multiplicand. Just remember that the numbers in the same diagonal row have a common place value and must be added to get the needed place value digit.

So, let's try another example, this time with some visual assistance with Figure 4.3.1-1. Putting the Sticks on a blank piece of paper and writing down the answers to the right, as shown, helps you see the step-by-step process.

Again, multiplication merely is collecting the right multiplicand Sticks and reading them right to the left for each place value number. For example, Figure 4.3.1-1 is 29 x6.

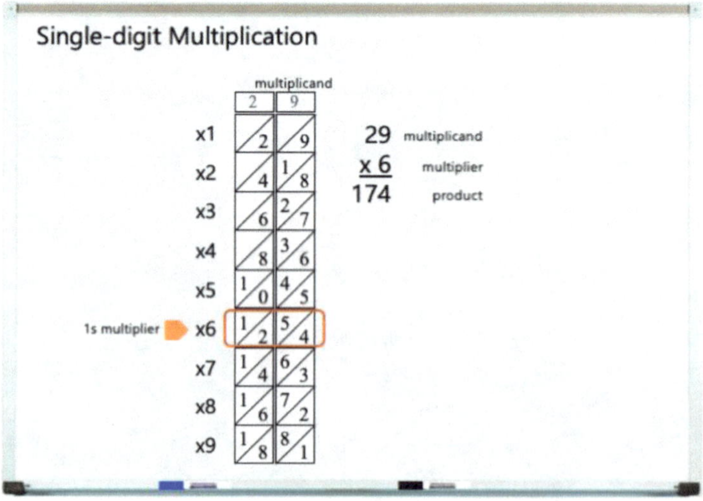

Figure 4.3.1-1: Tally-x Single-Digit Multiplier (SDM), 29 x6.

Collect the 2-stick and the 9-stick as shown. This is called the Multiplicand. Then look down to the 6th row, which is the single-digit Multiplier (x6). Read the numbers right to left across the row to get the Product. For example, there is a 4 in the 1s diagonal column. There is a 2 and 5 in the 10s diagonal column, which adds together for 7. Then, there is only a 1 in the 100s diagonal column. The Product is: 1, 2+5, 4 => 174.

Let's review this in more Place value detail. The 9-stick, by itself, is broken down with numbers in each row. All numbers in the lower diagonal position are in the 1s position because the 9-stick is used in the 1s position. All numbers in the high diagonal position are in the 10s position.

Why can't there also be a 100s position? This is one of the simple patterns of math, in that any single-digit number multiplied by the maximum single-digit number (9), cannot exceed a two-digit position, or a 10s place value. The maximum single-digit multiplicand is 9, and the maximum single-digit multiplier is 9. Therefore, 9 x9 is 81 by the 9-stick.

The 6th row yields the 10s diagonal value of 5 and the 1s value of 4; or 54. See the vertical Addition on the right in Figure 2.3 and the first value under the x6 line, which is 9+6.

The 2-stick approach represents the values of tens, so all numbers in the lower diagonal position are in the 10s position, and all numbers in the high diagonal position are in the 100s position. The 6th row yields the 100s value of 1 and the 10s value of 2; or 120. Again, notice the value 12 under the x6 line in Figure 4.3.1-1. This represents 120, where the zero is blank.

Combining the two stick values of the 6th row requires adding the proper Place values of 100s, 10s, and 1s. The Product is 1 of 100s, 2+5 of 10s, and 4 of 1s => 174.

This method works for any multiplicands as a collection of sticks, no matter how big the multiplicand value is.

Let's do a few more examples.

Single-Digit Multiplier (SSM)Tally-x Examples.

Example 1: 3 x8.

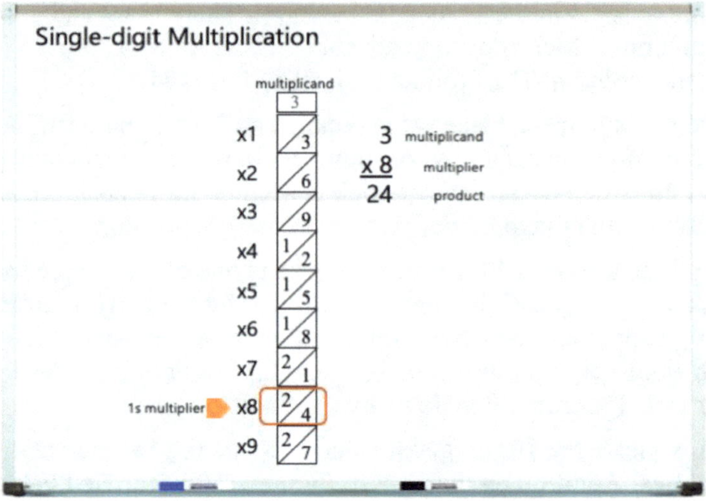

Figure 4.3.1-2: Multiplication of 3 x8.

The first thing we are going to do is a Single-digit multiplication problem. Starting with the problem eight (8-multiplier) times three (3-multiplicand), as seen in Figure 4.3.1-2 above, Notice the long-multiplication problem shown in the middle by the orange left pointer.

Note: The orientation of the numbers is traditionally the smallest to most significant; however, it does not matter. Go in the order that you are most comfortable with. In this case, the eight (8) allows us only to have to look down three (3) rows, to get the answer.

We are going to use the Tally-x stick number three (3), to represent the multiplicand. You will notice at the top of the Tally-x 3-stick is the number three (3), which matches the multiplicand in our problem.

The multiplier is eight (8), so I come down to the eighth (8th) row. This is shown by the 1x, 2x, 3x, 4x, 5x, 6x, 7x, and 8x orange right pointers. These represent the various row multipliers. Therefore, eight times three is 24, where 2 is the 10s place value, and 4 is the 1s place value. Altogether that makes the product 24.

It is essential to keep the problem number at the top, the multiplicand, the same as the number at the top of the Tally-x stick or sticks.

Example 2: 15 x4.

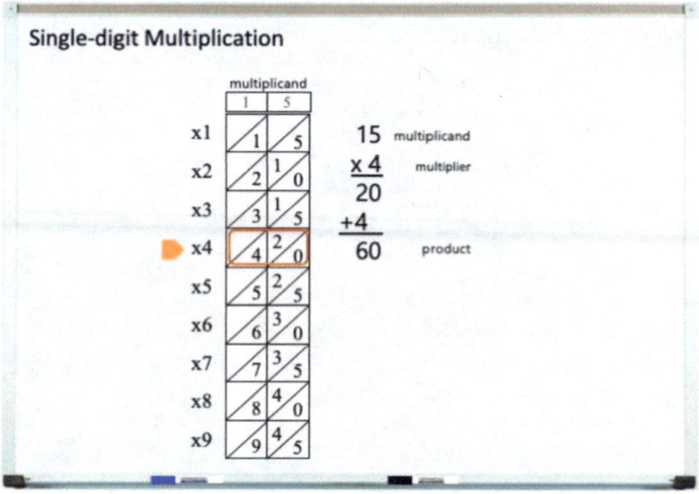

Figure 4.3.1-3: Multiplication of 15 x4.

In Figure 4.3.1-3, the first problem on the right is fifteen (15) times four (4). In the long multiplication layout that means the 15 is the multiplicand and 4 is the multiplier. Remember, the multiplicand represents the Tally-x sticks, so we need to build the value 15 using the 1-stick for the 10s placeholder and the 5-stick for the 1s placeholder, and put those to the left of the long multiplication problem.

The multiplier is a single-digit multiplier of four (4), so I come down to the fourth (4th) row and see the orange pointer. These represent the row multiplier. Therefore, four (4) times fifteen (15) is 0 in the 1s place value and 4+2 or 6 in the 10s place value. Altogether the product is 60.

It is less confusing to do this from the right, 1s place value, to the left, in this case, the 10s place value, so that if any place value is greater than 9, the resulting values can be carried over to the next higher place value. In this case that did not happen. This diagonal summation for each place value is the same as shifting the 10s place value for the double-digit (10s and 1s) value over one place value, or the 100s and 10s place values.

So far, we only need to know a single-digit addition, which is natural. As two signal digit numbers can never be more than eighteen (18) or 9 + 9, then finger counting is easy. However, you can also remove five from one or both numbers to get five (5) or ten (10) and then add that to the two remaining simple numbers. By the way, the last one is divided by five (5) or ten (10), and the remainder is then added.

Example 3: 45 x5.

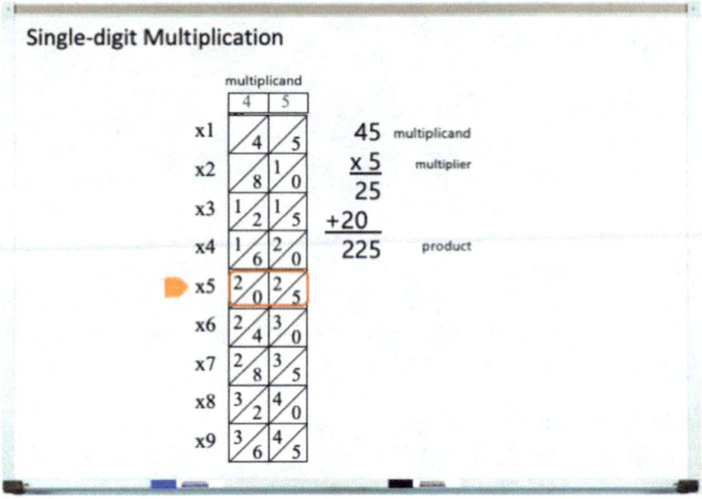

Figure 4.3.1-4: Multiplication of 45 x5.

Moving along to the third problem, see Figure 4.3.1-4; we have the problem forty-five (45) times five (5). In the long multiplication layout that means the 45 is the multiplicand and 5 is the multiplier.

The multiplicand represents the Tally-x sticks, so we need to build the value 45 using the 4-stick for the 10s placeholder and the 5-stick as the 1s placeholder, and put those to the left of the long multiplication problem.

The multiplier is a single-digit multiplier of five (5), so I come down to the fifth (5th) row. This is shown by the 1x, 2x, 3x, 4x, and 5x orange pointers. These represent the row multipliers. Therefore, five (5) times forty-five (45) is 5 is the 1s place value, 0+2 or 2 in the 10s place value, and 2 in the 100s place value. Altogether the product is 225.

As stated in the previous example, find the answer from the right or 1s place value and move to the next higher place value. In this example, that came in handy. To the left, in this case, is the 10s place value, so that if any place value is greater than 9, the resulting values can be carried over to the next higher place value.

That seems easy enough, but let's do a couple more single-digit multiplier problems using a 3-place and 4-place multiplicand.

Example 4: 144 x7.

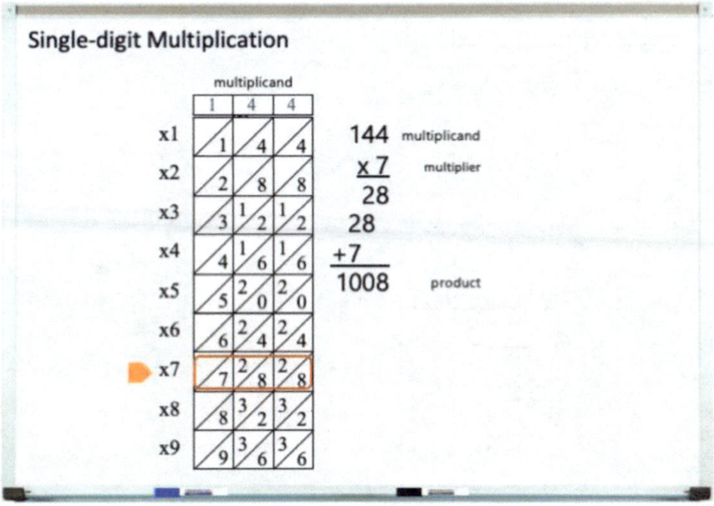

Figure 4.3.1-5: Multiplication of 144 x7.

The next problem on the right, top-down is shown in Figure 4.3.1-5 above. This is a 3-place multiplicand, and it can be that the problem is seven (7) times a hundred and forty-four (144). In the long multiplication layout, that means 144 is the multiplicand and 7 is the multiplier.

The multiplicand represents the Tally-x sticks, so we need to build the value 144 using the 1-stick for the 100s place holder, 4-stick for the 10s place holder, and another 4-stick as the 1s place holder; and put those to the left of the long multiplication problem, as shown.

The multiplier is a single-digit multiplier of seven (7), so I come down to the seventh (7th) row. This is shown by the 1x, 2x, 3x, 4x, 5x, 6x, and the 7x orange pointer, representing the single-digit row multipliers. Therefore, seven (7) times hundred and forty-four (144) is 8 in the 1s place value, 2 + 8 or 10 in the 10s place value, and 2+7 or 9 in the 100s place value. Altogether the product is 1008.

As stated in the previous examples, find the answer from the right or 1s place value and move to the next higher place value. In this example, that came in handy in this case the 10s place value being 10 or 0 with a carryover to the 100s of 1. In the 100s place value, we have 9, but we need to add the carry-over of +1. So, the 100s place value is no 10. We could stop there as there is no value in the 1000s place value. However, practice is best to follow the carryover of +1 to the 1000s place value.

Let's try a 4-place multiplicand next.

Example 5: 1844 x2.

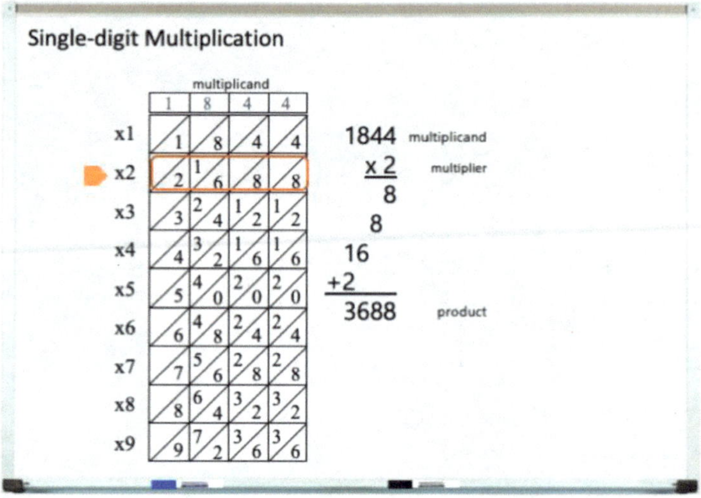

Figure 4.3.1-6: Multiplication of 1844 x2.

The last example shown in Figure 4.3.1-6 is a 4-place multiplicand, where the problem is two (2) times one-thousand, eight hundred and forty-four (1844). In the long multiplication layout, this means 1844 is the multiplicand and 2 is the multiplier.

To represent the multiplicand using Tally-x sticks, we build the value 1844: using the 1-stick for the 1000s place, the 8-stick for the 100s place, the 4-stick for the 10s place, and the 4-stick again for the 1s place. These are placed to the left of the long multiplication problem, as shown in the figure.

The multiplier is a single-digit multiplier of two (2), so we focus on the second row of sticks, indicated by the 2x orange pointer, which represents the single-digit row multiplier. Therefore, two (2) times one-thousand, eight hundred, and forty-four (1844) gives us 8 in the 1s place value, 8 in the 10s place value, 16 in the 100s place value, and 2 in the 1000s place value. Altogether, the product is 3,688.

4.3.2 No Carry Multiplication.

The greatest weakness of using 'carry values' above a multiplicand is the need to replace them in multi-digit Multipliers. As can be seen in the figure above, one could simply write down each multiplication effort in individual lines in long multiplication. The addition of the individual place values. But, in some cases, you have to use a carry value in the addition. Which again makes this difficult for multiplication of multi-digit multiplier.

To avoid this, the stick method has diagonal place values that can be added, thereby avoiding the carry marking. The addition of two single-digit additions is typically easy to do. In the next section of double and multi-digital multipliers, you will find the stick method very useful and helpful in future multiplications.

4.3.3 Double-Digit Multiplier (DDM).

Double-digit multiplication uses the same approach; you look for each multiplier digit (row) and write the value down to the right. In the problem in Figure 4.3.3-1 in the book, the 4 is a 10s digit, so the answer is 1160. Adding this to the case of the 6th row you get 1160 + 174 = 1334.

The method again uses the Place value system as learned in Pattern 11 and as we did in combining the place value in the single-digit multiplier. In this case, it might help to put the sticks on a blank piece of paper and write down the answers to the right, as shown in the figure. This will help you see the step-by-step process.

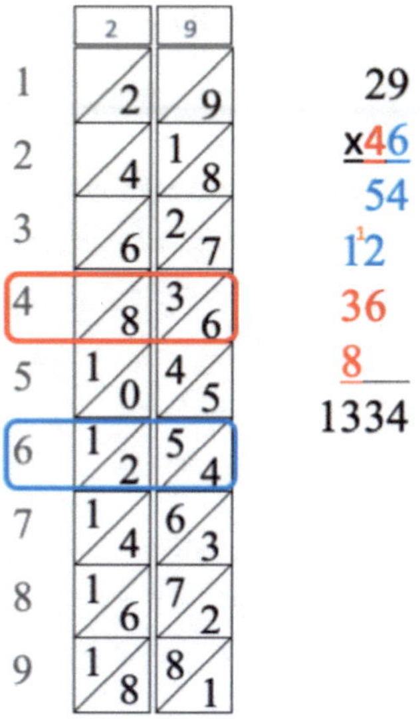

Figure 4.3.3-1: Tally-x Double Digit Multiplier Example.

With double-digit multiplication, you look for each multiplier digit (row) and write the value down to the right. In this case, the 4th row is a 10s digit which gives you 8+3 of 100s, and 6 of 10s => 1160. Do not forget that a 10s multiplier must have zero to the right to address the shift or blank position. The 6th row is the 1s digit, and that gives you 1 of 100s, 2+5 of 10s, and 4 of 1s => 174. Adding these two values together and you get 1160 + 174 = 1334.

This is the same process for any double-digit or any larger multiplier, treating each multiplier place value as a single-digit. Remember that you have to correct for each of the 10s, 100s, 1000s, or more multiplier place values. In the figure, we had a 4 in the 10s place value position, which means we see 40, or its Product of 116 is 1160. This is because the 10s place value position adds a zero to the Product's end.

This method works for any multiplier and is the same as the written-out long multiplication method you learned in school. The following are a few whiteboard examples.

4.3.4 Multi-Digit Multiplier Examples.

In the following multiplication section, only the final orange row pointer will be shown, and all place value diagonal summations will not be shown in the Figures.

Example 1: 3 x18.

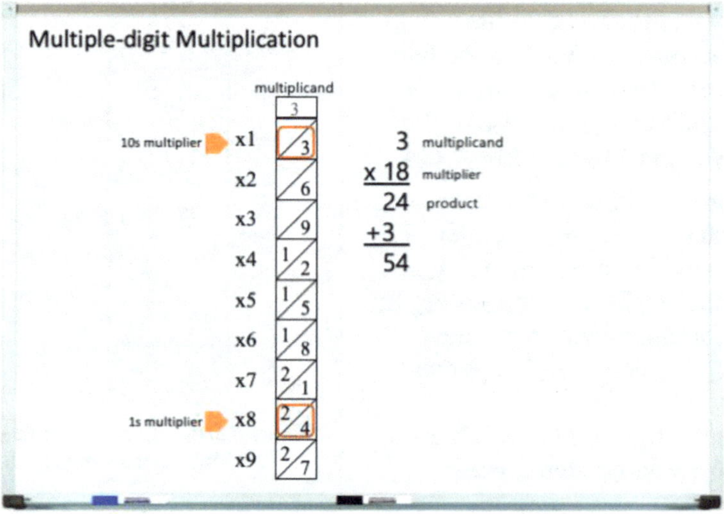

Figure 4.3.4-1: Multiplication of 3 x18.

The first thing we are going to do is a single-digit multiplication problem. We break the multiplier into individual single-digit place valves. So, let's start with problem eight (8) as the 1s place value multiplier times the three (3) multiplicand, as seen in Figure 4.3.4-1. Notice the long-multiplication problem shown in the middle by the orange left pointer.

Note: The orientation of the numbers is traditionally the smallest to most significant; however, it does not matter. Go in the order that you are most comfortable with. In this case, the eight (8) allows us only to have to look down eight (8) rows, to get the answer.

We are going to use the Tally-x stick number three (3), to represent the multiplicand. You will notice at the top of the 3-stick is number three (3), which matches the multiplicand in our problem.

The 1s multiplier is eight (8), so I come down to the eighth (8th) row. This is shown by the 8x orange pointer or row multiplier. Therefore, eight times three is 24, where 2 is the 10s place value, and 4 is the 1s place value. Altogether that makes the product 24. Combined with the one (1) times three (3) gives us 3 in the 10s place value or 30. Then, 24+30 = 54 is the answer.

It is essential to keep the problem number at the top, the multiplicand, the same as the number at the top of the Tally-x stick or sticks.

Example 2: 15 x14.

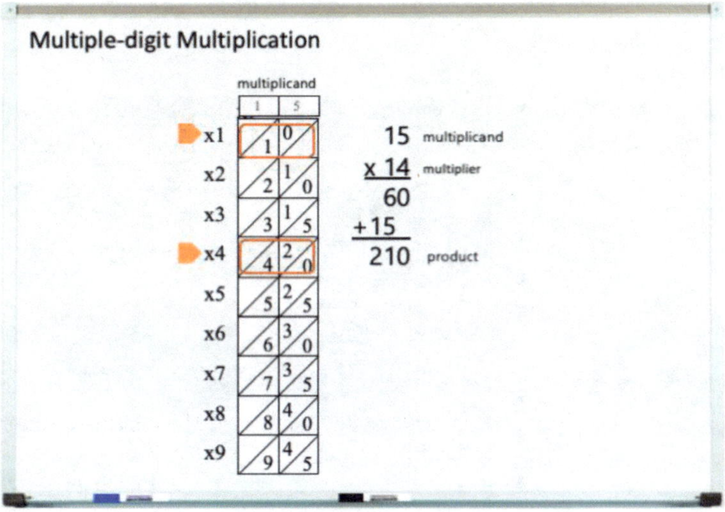

Figure 4.3.4-2: Multiplication of 15 x14.

Again, just as in the written Long Multiplication, we treat multi-digit multipliers as separate single-digit multipliers. So, in Figure 4.3.4-2, in this example, we have fourteen (14) as the multiplier times fifteen (15) as the multiplicand. We write down the long-multiplication problem in the middle with the multiplicand over the multiplier with a multiply bar drawn under the multiplier.

We are going to use the 1s stick to represent the multiplicand of one (1) in the 10s place value and the 5s stick to represent the five (5) in the 1s place value.

As we have two digits in the multiplier (14), we are going to individually do the 1s place value multiplier of four (4) and then the 10s place value multiplier of one (1), to follow the written long multiplication method.

1s multiplier: the multiplier for the 1s place value is the number four (4), so I come down the multiplicand counting stick (15) to the fourth row, indicated by the "4x" orange pointer. The answer is 0 in the 1s place value and 4+2 = 6 in the 10s place value or 60. This is written under the multiply bar and is 1s place value justified.

10s multiplier: the multiplier for the 10s place value is the number one (1), so I come down the multiplicand counting stick (15) to the first row, indicated by the "1x" orange pointer. The answer is 15, which is written under the 1s multiplier answer, but starting in the 10s place value or shifting left one place to get 150.

Now it is a matter of simple addition of adding 60 + 150 to get the product 210.

Example 3: 45 x25.

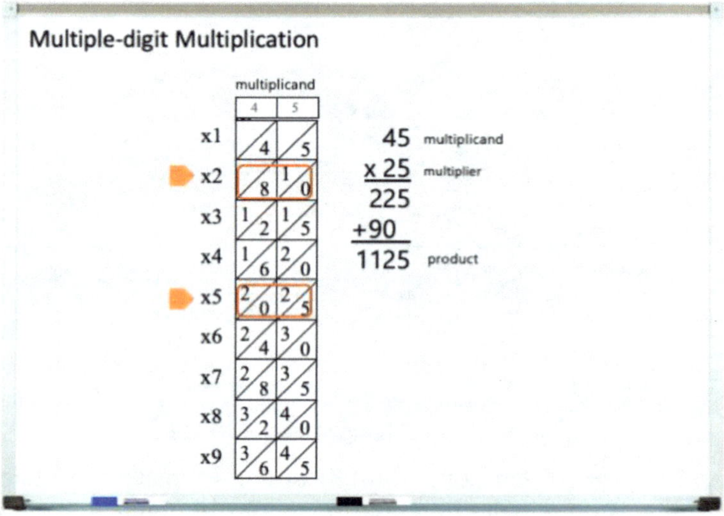

Figure 4.3.4-3: Multiplication of 45 x25.

In Figure 4.3.4-3, the problem is twenty-five (25) as the multiplier times forty-five (45) as the multiplicand. We write down the long-multiplication problem in the middle with the multiplicand over the multiplier with a multiply bar drawn under the multiplier.

We are going to use the 4s stick to represent the multiplicand of the 10s place value and the 5s stick to represent the five (5) in the 1s place value.

As we have two digits in the multiplier, we are going to individually do the 1s place value multiplier of five (5) and then the 10s place value multiplier of two (2), which follows the written long multiplication method.

1s multiplier: the multiplier for the 1s place value is the number five (5), so I come down the multiplicand stick of 45 to the fifth row, indicated by the

"5x" orange pointer. The answer is 5 in the 1s place value, 2 in the 10s place value (20), and 2 in the 100s place value (200), for a total of 225. This is then written under the multiply bar. It is justified by the 1s place value multiplier.

10s multiplier: the multiplier for the 10s place value is the number two (2), so I come down the multiplicand 15s stick to the second row, indicated by the "2x" orange pointer. The answer is 90, which is written under the 1s multiplier answer, but starting in the 10s place value, shift one.

Now it is a matter of simple addition, 900 + 225 to get the product of 1125.

Example 4: 144 x17.

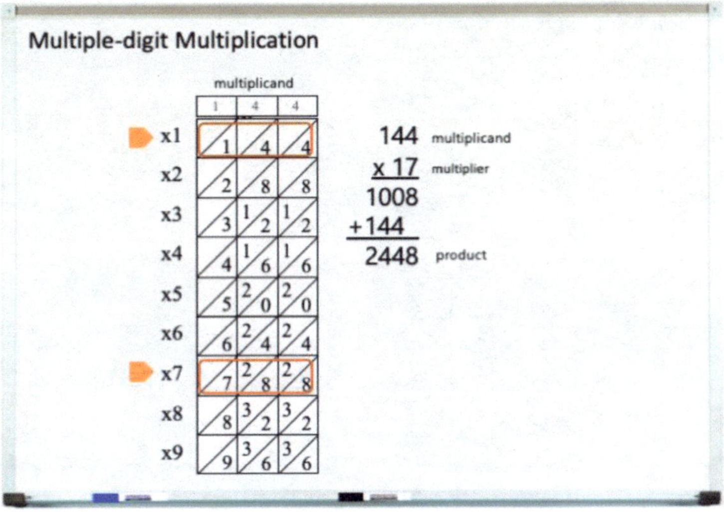

Figure 4.3.4-4: Multiplication of 144 x17.

In Figure 4.3.4-4, the problem is seventeen (17) as the multiplier times one hundred and forty-four (144) as the multiplicand. We write down the long-multiplication problem in the middle with the multiplicand over the multiplier with a multiply bar drawn under the multiplier.

We are going to use the 1s stick to represent the Multiplicand of one (1) in the 100s place value, the 4's stick to represent the four (4) in the 10s place value, and the 4's stick to represent the four (4) in the 1s place value or 144.

As we have two digits in the multiplier, we are going to individually do the 1s multiplier of seven (7) and then the 10s multiplier of one (1), which follows the written long multiplication method.

1s multiplier: the multiplier for the 1s place value is the number seven (7), so I come down the multiplicand 144 stick to the seventh row, indicated by the "7x" orange pointer. The answer is 8 in the 1's place value, 10 in the 10s place value, and 9 in the 100s place value and we get 1008, which is written under the multiply bar. It is justified by the 1s place value multiplier.

10s multiplier: the multiplier for the 10s place value is the number one (1), so I come down the multiplicand 144 stick to the first row, indicated by the "1x" orange pointer. The answer is 144, which is written under the 1s multiplier answer, but starting in the 10s place value or shift one.

Now it is a matter of simple addition 1008 + 1440 to get the product of 2448.

Example 5: 1844 x42.

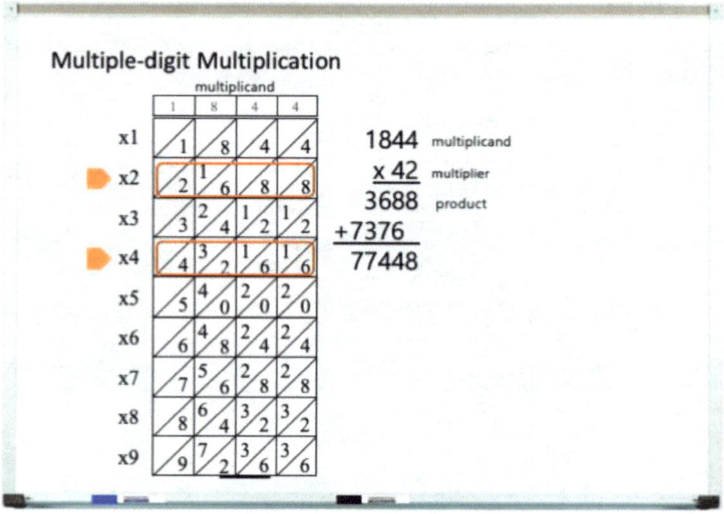

Figure 4.3.4-5: Multiplication of 1844 x42.

In Figure 4.3.4-5, the problem is forty-two (42) as the multiplier times one thousand, eight hundred, and forty-four (1844) as the multiplicand. We write down the long-multiplication problem in the middle with the multiplicand over the multiplier with a multiply bar drawn under the multiplier.

We are going to use the 1s stick to represent the Multiplicand of one (1) in the 1000s place value, the 8's stick to represent the eight (8) in the 100s place value and two 4s sticks to represent the four (4) in the 10s and 1s place values.

As we have two digits in the multiplier, we are going to individually do the 1s multiplier of two (2) and then the 10s multiplier of four (4), which follows the written long multiplication method.

1s multiplier: the multiplier for the 1s place value is the number two (2), so I come down the multiplicand 1844 stick to the second row, indicated by the "2x" orange pointer. The answer is 8 in the 1s place value, 8 in the 10s place value, 6 in the 100s place value, and 2 + 1 = 3 in the 1000s place value or 3688, which is written under the multiply bar in the 1s place value.

10s multiplier: the multiplier for the 10s place value is the number four (4), so I come down the multiplicand 144 stick to the fourth row, indicated by the “4x” orange pointer. The answer is 7376, which is written under the 1s multiplier, but starting in the 10s place value.

Now it is a matter of simple addition of 3688 + 73760 to get the product of 77448.

4.4: Multiplication Worksheets with Tally-x Sticks

4.4.1 Multiplication Worksheet 1: 1- and 1-digit examples.

Worksheet provided to help in the use and understanding of Tally-x multiplication. Traditionally, the multiplicand (stick) comes first, and then the multiplier (row value). Answers are in (…).

1.	3 x 7 = ______	(21)	26.	7 x 9 = ______	(63)
2.	8 x 6 = ______	(48)	27.	9 x 4 = ______	(36)
3.	5 x 2 = ______	(10)	28.	3 x 1 = ______	(3)
4.	7 x 9 = ______	(49)	29.	2 x 2 = ______	(4)
5.	1 x 2 = ______	(2)	30.	5 x 9 = ______	(45)
6.	5 x 7 = ______	(35)	31.	3 x 1 = ______	(3)
7.	2 x 8 = ______	(16)	32.	7 x 7 = ______	(49)
8.	7 x 2 = ______	(14)	33.	4 x 1 = ______	(4)
9.	4 x 9 = ______	(36)	34.	6 x 3 = ______	(18)
10.	5 x 8 = ______	(40)	35.	9 x 8 = ______	(72)
11.	2 x 4 = ______	(8)	36.	4 x 5 = ______	(20)
12.	4 x 3 = ______	(12)	37.	9 x 5 = ______	(45)
13.	8 x 9 = ______	(72)	38.	9 x 2 = ______	(18)
14.	3 x 4 = ______	(12)	39.	5 x 2 = ______	(10)
15.	8 x 4 = ______	(32)	40.	5 x 4 = ______	(20)
16.	6 x 1 = ______	(6)	41.	5 x 2 = ______	(10)
17.	1 x 9 = ______	(9)	42.	6 x 2 = ______	(12)
18.	4 x 3 = ______	(12)	43.	2 x 5 = ______	(10)
19.	8 x 7 = ______	(56)	44.	7 x 4 = ______	(28)
20.	7 x 3 = ______	(21)	45.	9 x 1 = ______	(9)
21.	7 x 2 = ______	(14)	46.	7 x 8 = ______	(56)
22.	5 x 2 = ______	(10)	47.	1 x 4 = ______	(4)
23.	3 x 7 = ______	(21)	48.	7 x 7 = ______	(49)
24.	1 x 2 = ______	(2)	49.	2 x 1 = ______	(2)
25.	8 x 6 = ______	(48)	50.	1 x 9 = ______	(9)

4.4.2 Multiplication Worksheet 2: 2- and 1-digit examples.

Worksheet examples to help in the use and understanding of Tally-x multiplication. In this sheet, the multiplicand is first (sticks) and then multiplier (row value). Answers are in parentheses.

1.	99 x 1 = _____	(99)	26.	25 x 2 = _____	(50)
2.	58 x 9 = _____	(522)	27.	64 x 2 = _____	(128)
3.	30 x 9 = _____	(270)	28.	85 x 8 = _____	(680)
4.	22 x 5 = _____	(110)	29.	89 x 4 = _____	(356)
5.	44 x 4 = _____	(176)	30.	42 x 9 = _____	(378)
6.	50 x 4 = _____	(200)	31.	96 x 2 = _____	(192)
7.	91 x 9 = _____	(819)	32.	21 x 6 = _____	(126)
8.	47 x 2 = _____	(94)	33.	74 x 2 = _____	(148)
9.	86 x 9 = _____	(774)	34.	87 x 5 = _____	(435)
10.	60 x 6 = _____	(360)	35.	60 x 7 = _____	(420)
11.	55 x 2 = _____	(110)	36.	34 x 9 = _____	(306)
12.	34 x 2 = _____	(68)	37.	48 x 5 = _____	(240)
13.	45 x 4 = _____	(180)	38.	74 x 7 = _____	(518)
14.	51 x 1 = _____	(51)	39.	53 x 5 = _____	(265)
15.	86 x 7 = _____	(602)	40.	32 x 6 = _____	(192)
16.	47 x 3 = _____	(141)	41.	63 x 9 = _____	(567)
17.	63 x 3 = _____	(189)	42.	26 x 1 = _____	(26)
18.	43 x 8 = _____	(344)	43.	19 x 8 = _____	(152)
19.	62 x 1 = _____	(62)	44.	22 x 3 = _____	(66)
20.	16 x 6 = _____	(96)	45.	17 x 7 = _____	(119)
21.	38 x 9 = _____	(342)	46.	96 x 5 = _____	(480)
22.	66 x 5 = _____	(330)	47.	97 x 8 = _____	(776)
23.	68 x 5 = _____	(340)	48.	90 x 9 = _____	(810)
24.	11 x 5 = _____	(55)	49.	81 x 8 = _____	(648)
25.	26 x 8 = _____	(208)	50.	58 x 4 = _____	(232)

4.4.3 Multiplication Worksheet 3: n- and 1-digit examples.

Worksheet examples to help in the use and understanding of Tally-x multiplication problems. the multiplicand (sticks) is first, and then the multiplier (row value). Answers are in (…).

1.	1,083 x 3 = _____	(3,249)
2.	507 x 9 = _____	(4,563)
3.	5,937 x 9 = _____	(53,433)
4.	65 x 4 = _____	(260)
5.	7,669 x 8 = _____	(61,352)
6.	8,045 x 5 = _____	(40,225)
7.	780 x 7 = _____	(68,460)
8.	5,213 x 4 = _____	(20,852)
9.	449 x 8 = _____	(3,592)
10.	467 x 6 = _____	(2,802)
11.	4,855 x 6 = _____	(29,130)
12.	1,818 x 4 = _____	(7,272)
13.	518 x 6 = _____	(31,110)
14.	4,953 x 3 = _____	(14,859)
15.	97 x 4 = _____	(388)
16.	1,193 x 5 = _____	(5,965)
17.	98 x 8 = _____	(31,880)
18.	2,030 x 2 = _____	(4,060)
19.	6,149 x 1 = _____	(6,149)
20.	54 x 5 = _____	(27,265)
21.	389 x 7 = _____	(51,723)
22.	2,994 x 9 = _____	(26,946)
23.	27 x 6 = _____	(39,762)
24.	2,941 x 1 = _____	(2,941)
25.	3,023 x 6= _____	(18,138)
26.	3,048 x 9 = _____	(27,432)
27.	3,778 x 3 = _____	(11,334)
28.	8,973 x 5 = _____	(44,865)
29.	661 x 1 = _____	(661)
30.	1,227 x 2 = _____	(2,454)
31.	44 x 4 = _____	(176)
32.	4,138 x 6 = _____	(24,828)
33.	502 x 2 = _____	(1,004)
34.	7,635 x 5 = _____	(38,175)
35.	3,294 x 7 = _____	(23,058)
36.	198 x 3 = _____	(594)
37.	5,251 x 6 = _____	(31,506)
38.	5,340 x 3 = _____	(16,020)
39.	915 x 8 = _____	(7,320)
40.	82 x 2 = _____	(164)
41.	8,915 x 5 = _____	(44,575)
42.	88 x 9 = _____	(792)
43.	6,751 x 4 = _____	(27,004)
44.	2,285 x 1 = _____	(2,285)
45.	7,383 x 6 = _____	(44,298)
46.	1,285 x 3 = _____	(3,855)
47.	77 x 7 = _____	(539)
48.	2,916 x 2 = _____	(5,832)
49.	3,298 x 4 = _____	(13,192)
50.	383 x 8 = _____	(3,064)

4.4.4 Multiplication Worksheet 4: n- and 2-digit examples.

Worksheet examples to help in the use and understanding of Tally-x multiplier problems. In this sheet, the multiplicand (sticks) comes first, and then the multiplier (row value). Note how Pattern 4 can be used here to show that all the multiplicand values with four digits with a 2-digit multiplier results in either a 5- or 6-digit answer. Answers are in (…).

1. 1,693 x 93 = _____ (157,449)
2. 8,055 x 29 = _____ (233,595)
3. 4,034 x 79 = _____ (318,686)
4. 8,480 x 21 = _____ (178,080)
5. 4,513 x 88 = _____ (397,144)
6. 5,017 x 35 = _____ (175,595)
7. 7,158 x 67 = _____ (479,586)
8. 2,938 x 14 = _____ (41,132)
9. 8,626 x 48 = _____ (414,048)
10. 5,002 x 96 = _____ (480,192)
11. 1,060 x 16 = _____ (16,960)
12. 8,105 x 24 = _____ (194,520)
13. 1,408 x 86 = _____ (121,088)
14. 8,282 x 33 = _____ (273,306)
15. 918 x 64 = _____ (58,752)
16. 2,906 x 15 = _____ (43,590)
17. 6,414 x 78 = _____ (500,292)
18. 1,330 x 22 = _____ (29,260)
19. 5,568 x 52 = _____ (289,536)
20. 5,397 x 90 = _____ (485,730)
21. 6,021 x 53 = _____ (319,113)
22. 4,228 x 57 = _____ (240,996)
23. 2,444 x 89 = _____ (217,516)
24. 2,996 x 31 = _____ (92,876)
25. 3,729 x 72 = _____ (268,488)
26. 894 x 10 = _____ (8,940)
27. 6,701 x 11 = _____ (73,711)
28. 7,789 x 74 = _____ (576,386)
29. 9,918 x 32 = _____ (317,376)
30. 1,694 x 81 = _____ (137,214)
31. 4,177 x 84 = _____ (350,868)
32. 3,027 x 44 = _____ (133,188)
33. 6,603 x 19 = _____ (125,457)
34. 5,910 x 61 = _____ (360,510)
35. 3,039 x 37 = _____ (112,443)
36. 5,973 x 97 = _____ (579,381)
37. 4,941 x 59 = _____ (291,519)
38. 9,370 x 68 = _____ (637,160)
39. 4,382 x 51 = _____ (223,482)
40. 5,616 x 73 = _____ (409,968)
41. 7,360 x 46 = _____ (338,560)
42. 9,800 x 85 = _____ (833,000)
43. 3,817 x 36 = _____ (137,412)
44. 3,241 x 30 = _____ (97,230)
45. 1,219 x 42 = _____ (51,198)
46. 32,759 x 51 = _____ (1,670,709)
47. 4,575 x 92 = _____ (420,808)
48. 7,801 x 70 = _____ (546,070)
49. 253 x 20 = _____ (5,060)
50. 13,654 x 28 = _____ (382,312)

4.5: Tally-x Multiplication Cutout.

The figure in the book can be folded, glued on the back, folder together, cutting it out; keeping the columns aligned front to back, then cutting the aligned edges. Sticks can be cut vertically.

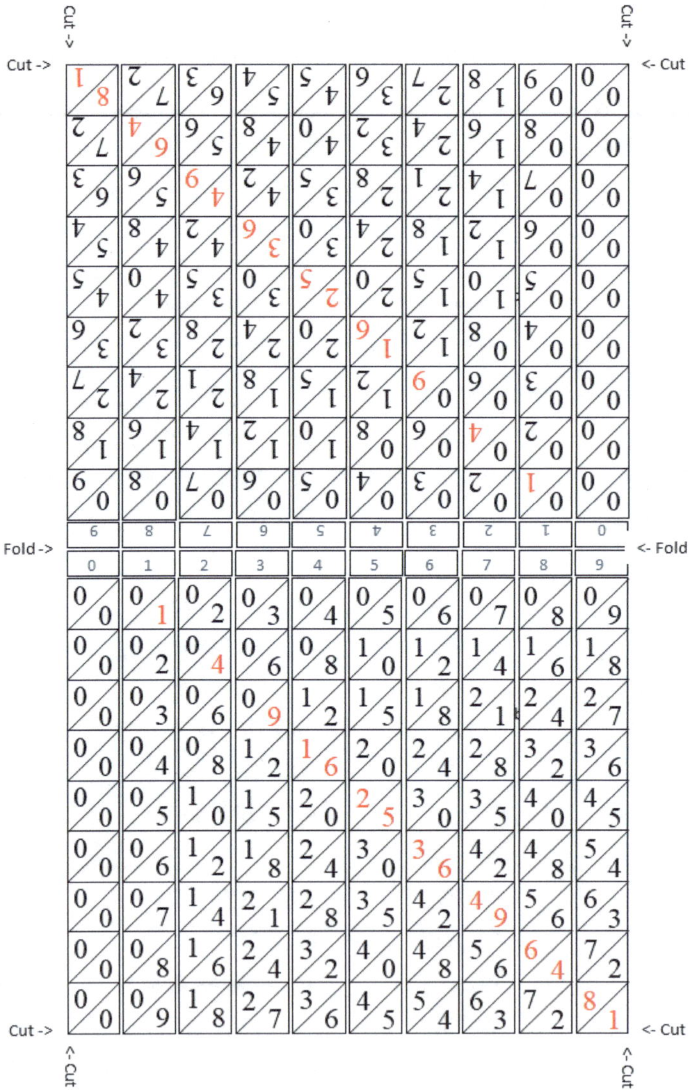

Figure 4.5-1: Tally-x Multiplication Cutout.

Chapter 5: Division

0	1	2	3	4	5	6	7	8	9
0/0	0/1	0/2	0/3	0/4	0/5	0/6	0/7	0/8	0/9
0/0	0/2	0/4	0/6	0/8	1/0	1/2	1/4	1/6	1/8
0/0	0/3	0/6	0/9	1/2	1/5	1/8	2/1	2/4	2/7
0/0	0/4	0/8	1/2	1/6	2/0	2/4	2/8	3/2	3/6
0/0	0/5	1/0	1/5	2/0	2/5	3/0	3/5	4/0	4/5
0/0	0/6	1/2	1/8	2/4	3/0	3/6	4/2	4/8	5/4
0/0	0/7	1/4	2/1	2/8	3/5	4/2	4/9	5/6	6/3
0/0	0/8	1/6	2/4	3/2	4/0	4/8	5/6	6/4	7/2
0/0	0/9	1/8	2/7	3/6	4/5	5/4	6/3	7/2	8/1

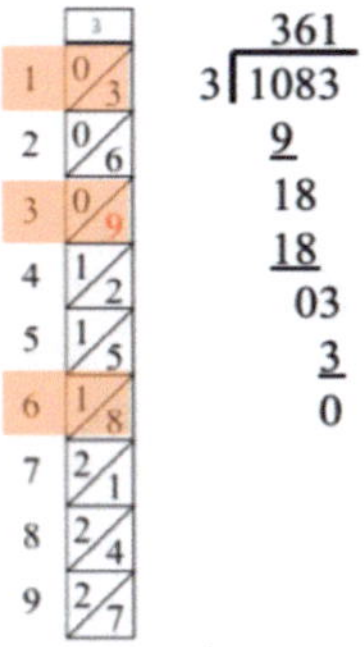

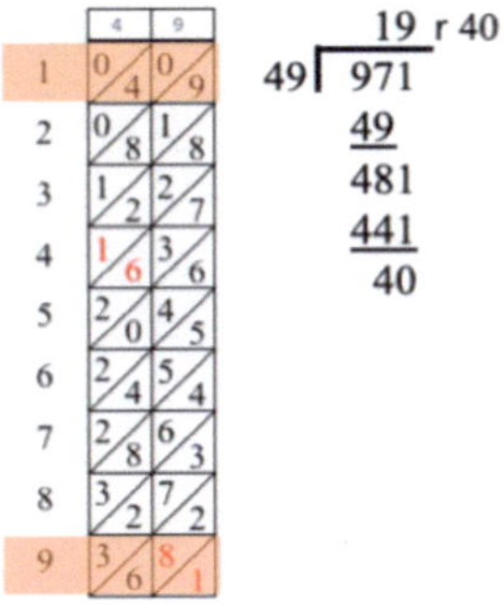

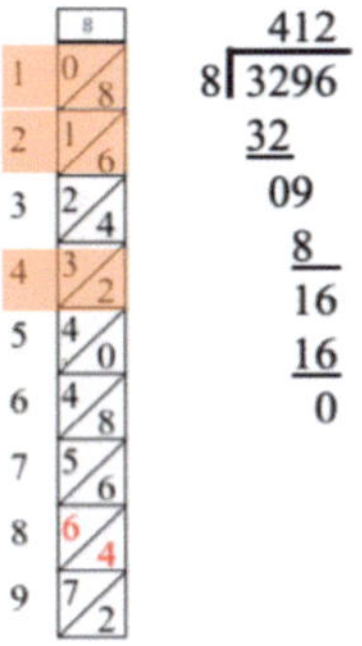

5.1: Basics of Division.

Division with Tally-x is more natural than the multiplication method because you do not have to use the place value system as was demonstrated in the 2-digit or larger multiplier method in the previous section. As pointed out in Figure 1.4 -1, the roadmap of arithmetic, just as multiplication is the repeating pattern or count of the same number value added together. So is division the repeating count of the same number value subtracted from a value. But, a shortcut simplifies that count using the same sticks that represent the divisor, see Figure 5.1-1.

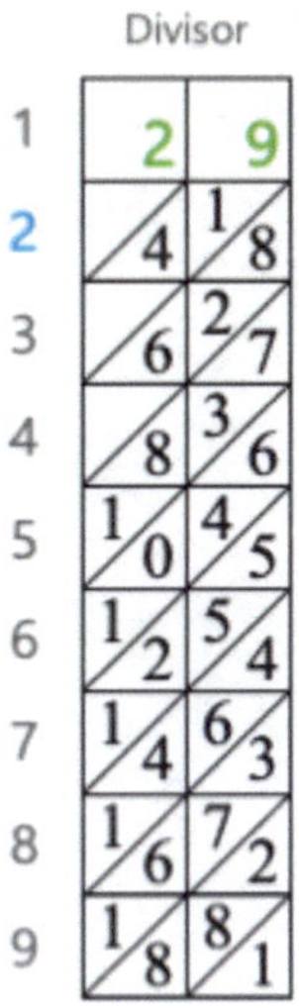

Figure 5.1-1: Tally-x Division Example of 601 ÷29.

Division (÷) is merely collecting the correct Sticks and looking down the row for the first value that is the same or smaller than the first set of values in the Dividend. Figure 5.1-2 shows the names of each of the four division components: Divisor, Quotient, Dividend, and Remainder.

The rules of Divison are simple, count the place values in the Divisor. Then use that count for the Dividend, and look down the Divisor stick for a value equal to just less than that value.

For example, in Figure 5.1-1, the divisor has a place value count of two (2), so from the dividend (601), 60 is the value to search. All you need to do is look down the 29 stick rows till you find a number less than 60. This is the 2nd row or 58. Subtract the 58 from what you want (60), and you have 2. See Figure 5.1-2. Bring down the next place value (1), and you have 21. Since 20 is smaller than 29, it is a final reminder. The Quotient is 20 with the remainder of 21.

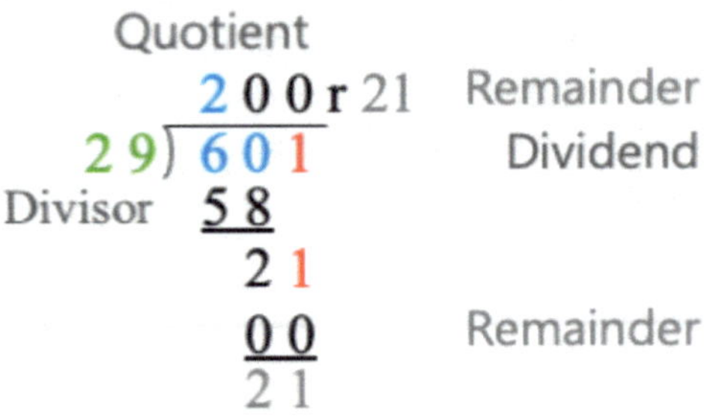

Figure 5.1-2: Dividend divided by Divisor = Quotient and Remainder.

Option: you can go further but that requires making the Divisor a decimal value and that is covered in Chapter 6.

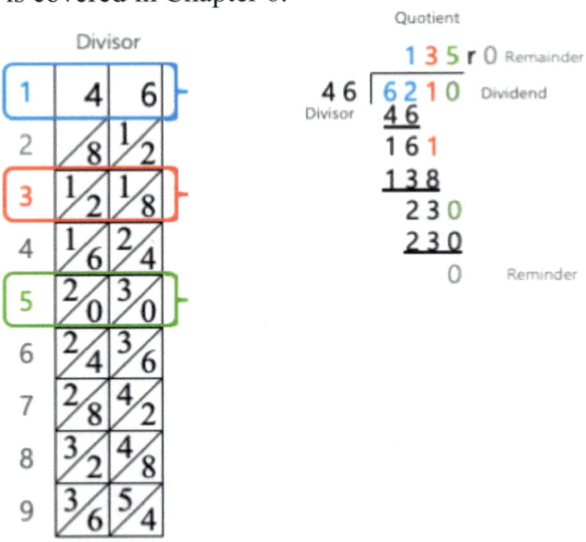

Figure 5.1-3: Tally-x Division Example of 6210 ÷46.

In the next, Figure 5.1-3, we divide 6,210 by 46. Therefore, we use the 4 and 6 sticks to create the divisor 46. Since 62 is larger than or equal to the divisor 46, we find that the first quotient value is the 1st row. Subtracting 46 from what you want (62) and the left with 16. See the long division to the right in Figure 5.1-3

Bring down the 1 from the dividend to get 161. The 3rd-row value (138) gives the next quotient value of 3 or the 10s place value of the quotient. Subtract 138 from 161 and that leaves 23.

Now finally bring down the 0 from the dividend to get 230. The 5th-row value (230) from the stick gives the next quotient value of 5 for the 1s place value. Subtracting 230 from what you want, or 230, and you are left with 0. In this case, there is no remainder in this problem.

So, the Quotient is the combination of the 100s, 10s, and 1s place values, which yields 135 and no remainder.

This method works for any divisor, whether it is a single-, double-, or larger digit. This is why it is easier than multiplication. It is also the same as the written-out method, Long Division, you will learn in school, as seen on the right side of Figure 5.1-3. Try it yourself, however, do remember the place value system when writing down your answer under the dividend.

Division, as learned by the Tally-x Stick method, will seem easier than multiplication, as one only needs to create the divisor on the sticks and count the place values of the divisor. More Stick examples will follow in section 5.3.

But, next is the Traditional Long Division for clarity.

5.2: Traditional Long Division.

Just as in Multiplication (Section 4.2: Traditional Multiplication) there is both a short and long division used for the vertically written calculations. The short division method is used for one-digit numbers only, and long division is used for two- or more-digit numbers that involve carry-over values to the next higher place values. Both use the same identifiers: Dividend, Divisor, and quotient, but the division has a remainder for non-perfect dividends. Non-perfect Dividends are those values that fall between the multiplications shown in the multiplication table or the Tally-x sticks.

Divisor: the number that the dividend will be divided by, to the left.

- Guess: traditionally this is what the student must do to assess the quotient.
- Dividend: the number that is being divided, under the division bar symbol.
- Quotient: the division solution, over the division bar symbol.
- Remainder: a number that is less than the divisor and leftover, indicated by a small 'r'.

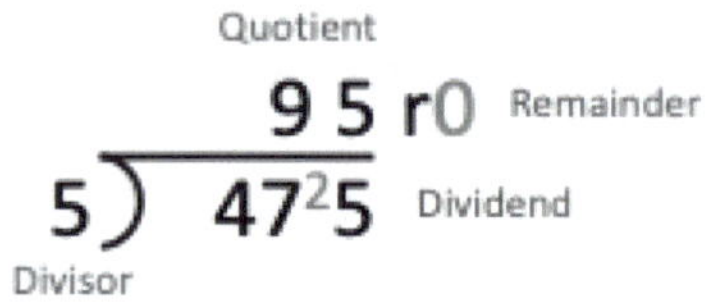

Figure 5.2-1: 475 ÷5.

5.2.1: Short Division.

In written calculation, the short division method is used for dividing by one-digit numbers only. This method requires more mental subtractions, where the difference between the quotient times the divisor is less than the applicable dividend. That difference is then written above and to the right of the applicable dividend. This raised number combined with the next applicable dividend value becomes the new dividend. See Figure 5.2.1-1 for a quick example.

Example 1: 847 ÷8.

In this case, the dividend is eight-hundred and forty-seven (847), and the divisor is five (5), so the dividend is written with the division brace over it and to the left of the dividend. Then the divisor is written to the left of the brace. Once written correctly, the process is just a matter of dividing the divisor into the first few digits of the dividend. The division is finding how many times the divisor can fit into the applicable dividend and not be larger. This is where the memorization of the multiplication table is traditionally used. This is the "Guessing" part of the instructions.

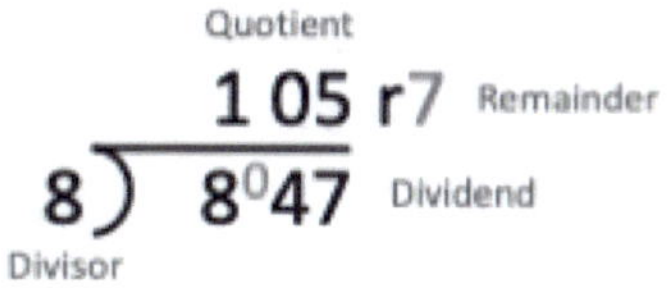

Figure 5.2.1-1: 484775 ÷8.

In this example, how many times 8 can be fitted into 8 or 84 if 8 is smaller than the divisor? In this case, it is just how many times 8 can be fitted into 8. Well, 1 x8 is 8, and 2 x8 is 16. The latter is larger than 8, so the quotient is 1 from the multiplication of 1 x8. So, a 1 is written over the 8 as the 100s place value quotient, and the mentally multiplied value is subtracted 1 x8 = 8 from the dividend of 8, or 0. The 0 is then written between the 8 and 4 of the dividends.

The new quotient is 4, but it is too small for the divisor 8, so the 10s place value product is 0.

We now have a new dividend of 47 depending. So, how many times 8 can be fitted into 47? Well, per the Rote learning method: 1 x8 = 8, 2 x8 = 16, 3 x8 = 24, 4 x8 = 32, 5 x8 = 40, 6 x8 = 48, 7 x7 is 56, etc. The sixth answer is greater than 47 by 1, so the new quotient is 5. The difference between the divisor times the new quotient, 5 x8 = 40, and the new dividend is 7. The new quotient 5 is written over the 7 part of 7, so the new quotient is now 105, and the new dividend is 7. This is the remainder and is written to the right of the quotient, preceded with 'r'.

This is a non-perfect division because it has a remainder, of 7 so the answer to how many times 8 can fit into 847 is 105 times with a remainder of 7.

Example 2: 97 ÷4.

In this case, the dividend is ninety-seven (97), and the divisor is four (4), so the dividend is written with the division brace over it and to the left of the dividend. Then the divisor is written to the left of the brace. Once written correctly, the process is just a matter of dividing the divisor into the first few digits of the dividend. The division is finding how many times the divisor can fit into the applicable dividend and not be larger. This is where the memorization of the multiplication table or Rote learning is used.

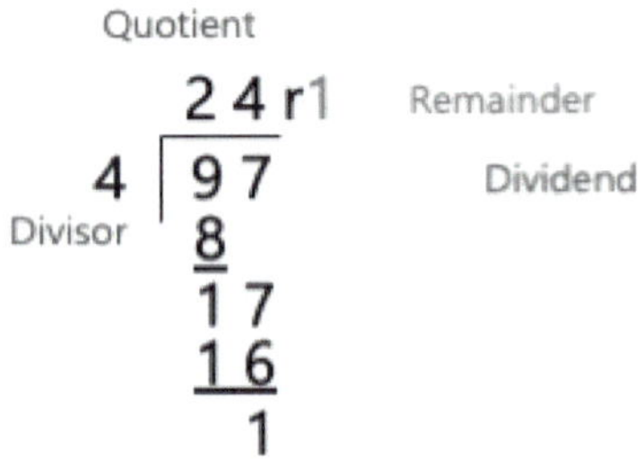

Figure 5.2.1-2: 97 ÷4.

In this example, how many times can 4 be fitted into 9 or 97 if 9 is smaller than the divisor? In this case, it is just how many times can 4 be fitted into 9. Well, 1 x4 = 4, 2 x4 = 8, 3 x4 is 12, etc. The 3 x4 = 12 is bigger than 9, so the quotient is 2 from the multiplication of 2 x4 = 8. So, a 2 is written over the 9, which is the 10s place value, and the difference is subtracted from 9 minus 8 = 1 is then written between the 9 and 7 of the dividends. This is done to show the next dividend is 17.

We now have a new dividend of 17. So, how many times can 4 be fitted into 17? Well, 1 x4 = 4, 2 x4 = 8, 3 x4 = 12, 4 x4 = 16, 5 x4 = 20, etc. The 5 x4 = 20 one is greater than 17, so the new quotient is 4. This 4 is written over the 7, for a quotient of 24. The difference between the divisor and the new quotient, 4 x4 = 16, and the new dividend (17) is 1. This is the remainder and is written to the right of the quotient, preceded by an r.

This is a non-perfect division because it has a remainder, so the answer to the question of how many times 4 can fit into 97 is 24 times plus 1.

Overall, the process seems similar to the steps in long division, but here you don't have to bring any numbers down, subtract, or multiply. Besides, it takes up less space on paper. However, it does require rigorous memorization of the multiplication tables so that division problems can be solved in the head, or by use of the Rote learning method.

5.2.2: Long Division

The method for teaching long division in written form eliminates the mental subtraction process as each approximation of the answer is written beneath the applicable dividend. This method is also used for giving answers with remainders, fractions, and decimals. See Figure 5.2.1-1 for a quick overview of 127 divided by 4.

Example 1: 127 ÷4.

In this case, the dividend is one-hundred and twenty-seven (127), and the divisor is four (4), so the dividend is written with the division brace over it and to the left of the dividend. Then the divisor is written to the left of the brace. Once written correctly, the process is just a matter of dividing the divisor into the first few digits of the dividend. The division is finding how many times the divisor can fit into the applicable dividend and not be larger. This is where the memorization of the multiplication table is used. This is the "Guessing" part of the instructions.

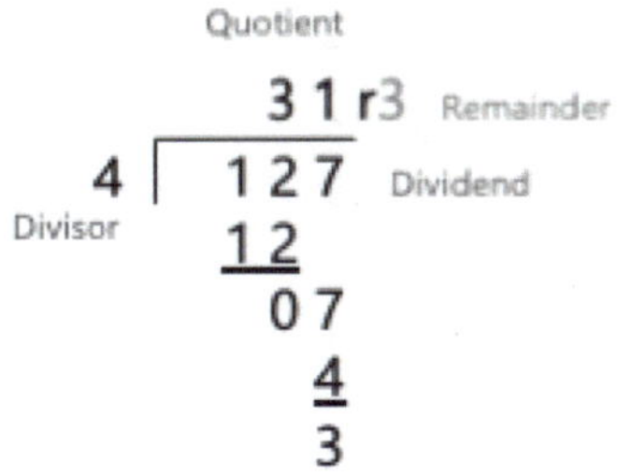

Figure 5.2.2-1: 127 ÷4.

In this example, how many times can 4 be fitted into 1 or 12 if 1 is smaller than the divisor? In this case, the divisor is larger than 1, so the appropriate dividend is 12. Well, 1 x4 = 4, 2 x4 = 8, 3 x4 = 12, 4 x4 = 16, etc. This is where 4 x4 = 16 is larger than 12, so the quotient is 3 from the multiplication of 3 x4 = 12. So, a 3 is written over the 12 as a 10s place value quotient, and the multiplied value 3 x4=12 is written below the appropriate dividend (12). The difference between these two 12-values is zero (0). We then bring down the next dividend value (7).

With no 10s place value, a difference of 0, we now have a new dividend of 7. So, how many times can 4 be fitted into 7? Well, 1 x4 = 4 and 2 x4 = 8, and the last one is greater than 7, so the new quotient is 1. The difference between the divisor and the new quotient is 1 x4 = 4, and the new dividend is 3. This 3 is written as the remainder to the right of the quotient, proceeded with an r.

This is a non-perfect division because it has a remainder that is not zero (0), so the answer to the question is that 4 can fit into 127, 31 times plus a remainder of 3.

Example 2: 6257 ÷37.

In this example, the dividend is six thousand, two-hundred, and fifty-seven (6257), and the divisor is thirty-seven (37). Just as above, this is written so written with the division brace over the dividend. Then the divisor is written to the left of the brace.

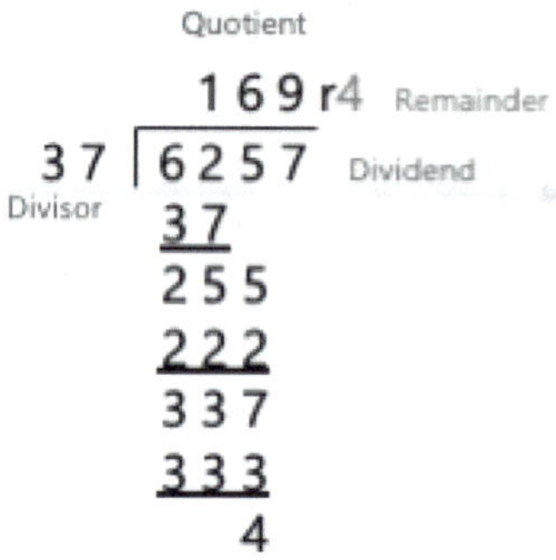

Figure 5.2.2-2: 6257 ÷37.

As a reminder, Division is finding how many times the divisor can fit into the applicable dividend and not be larger. This is where the memorization or Guessing of the multiplication table is used.

In this example, how many times can 37 be fitted into 6 or 62 if 37 is smaller than the divisor? In this case, it is the latter, as the divisor is larger than 6. So, how many times can 37 be fitted into 62? Well, 1 x37 = 37 and 2 x37 = 74. The latter is larger than 37, so the quotient is 1 from the multiplication of 1 x37. The 1 is written over the 62 in the 100s place value of the quotient, and then 1 x37 = 37 value is subtracted from 62 to leave 25. This is then written under the 37. The next dividend value (5) is dropped down to form 255. See Figure 5.2.2-2.

We now have a new dividend of either 255 or 2557 depending on if the first is smaller than the divisor. In this case, 255 is not. So, how many times can 37 be fitted into 255? Well, 1 x37 = 37, 2 x37 = 74, 3 x37 = 111, 4 x37 = 148, 5 x37 = 185, 6 x37 = 222, 7 x37 = 259, etc. The last one is greater than 255, so the new quotient is 6 and it is written over the dividend of 5 as the 10s place value of the quotient. The difference between the new quotient, 6 x37 = 222, and the new dividend (255) is 33. This 33 is written under the 222 value, right justified. The last dividend value (7) is dropped down to become 337.

We now have a new dividend of 337. So, how many times can 37 be fitted into 337? Well, 1 x37 = 37, 2 x37 = 10, 3 x37 = 15, 4 x37 = 20, 5 x37 = 185, 6 x37 = 222, 7 x37 = 259, 8 x37 = 296, and 9 x37 is 333. The difference between the new quotient, 9 x37 = 333, and the new dividend (337) is 4. This 4 is now the remainder.

As in the Tally-x sticks multiplicand, you can't go higher than 9 times the divisor. If you do, that means you got the last quotient value wrong. The 9 is written over the dividend 7 as the quotient's 1s place value. This is a non-perfect division so the remainder is written to the right of the quotient proceeded with an r. The answer is 169 r4.

The advantage of this method is that the method approach is clearly shown and makes the repetitive process easier to remember. However, it also requires the memorization of the multiplication table or the Rote method. Because it is clearly shown, it is best to show your teacher you understand the division process.

5.3: Tally-x Stick Division

In this section, we are going to do single-digit divisor problems and show you the easy way to do long division by examples. In each example, the problems are listed on the right, top-down. Remember from section 4.2 that the Dividend is under the division brace and the Divisor is outside the division brace.

5.3.1: Single-digit Divisor (SDD).

Example 1: 24 ÷8.

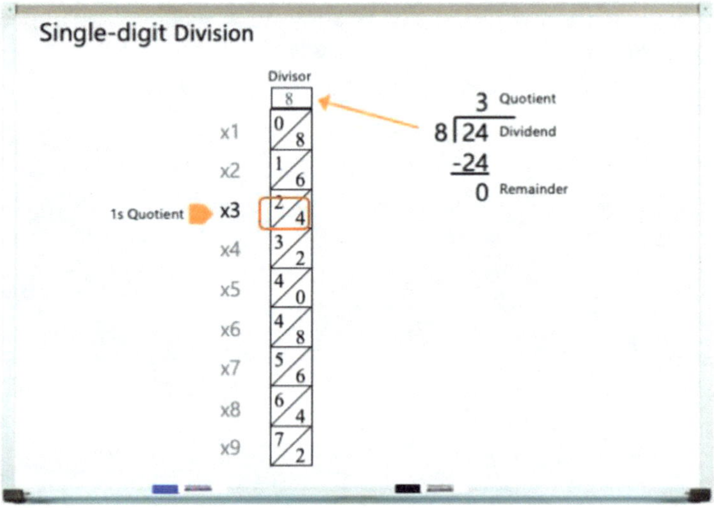

Figure 5.3.1-1: Whiteboard Division: 24 ÷8.

This first example is a single-digit division problem. Starting with the problem, twenty-four (24) dividends are divided by the eight (8) divisor, as seen in before in Figure 3.3.1-2.

The long-division setup is similar to Figure 5.1.1-1, shown where the divisor is 8, and the dividend 24 is under the division bar. The quotient and remainder are yet to be found.

The first step is to create a Tally-x stick version of the divisor, see Figure 5.3.1-1, which is 8 as seen on the left of the whiteboard. In traditional division, you would “guess” how many times the divisor can go into the dividend or how many times the quotient is closest to the dividend. Using the sticks for the divisor, we eliminate the "guessing." Simply look down the rows of the stick to find a number that is equal to or just smaller than the dividend.

Before proceeding, identify which digits in the dividend to use. Firstly, recall or follow Pattern 11 for division. We have a 2-digit dividend and a 1-digit divisor.

Question: Is the 1st digit of the dividend, 2, larger than the divisor (8), which is in the first row on the stick?

Since it is smaller, proceed to the 2nd digit of the dividend and check its size.

If the new 2-digit dividend is smaller, continue adding a new digit until it is not.

Now that we have determined the correct number of digits to work with, look down the stick, rows 1 through 9, to find a number that is equal to or just smaller than the dividend. In this case, we have 24 on the stick at row 3, which equals our dividend of 24.

Place the 3 over the last digit of the dividend, which is 4. Place the answer, 8 × 3 = 24, from the stick under the dividend of 24. Now subtract and write the solution under the answer of 8 × 3.

If the subtraction results in a remainder greater than 0, restart at step 1 with the remainder, plus the next dividend value, even if it is zero (0).

The answer is 3 with no remainder.

Example 2: 25 ÷8.

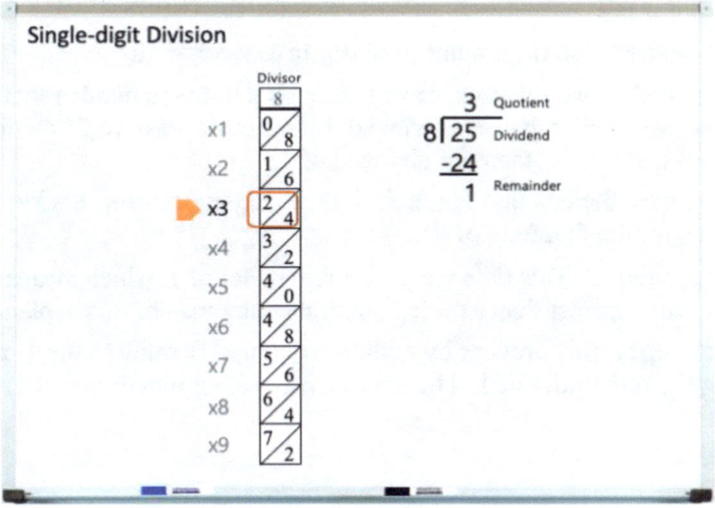

Figure 5.3.1-2: Whiteboard Division: 25 ÷8.

Not all division problems have a remainder of 0 as in example 1. Just as not all products in the store are sold in dollar ($) values and cost only dollars at the checkout stand. In example 2, think of the remainder as cents on the dollar.

So, let's change our problem so that it is eight into 25. The long-division set-up is still the same, where the divisor is still 8, and the dividend is now 25, which is one more than perfect, under the division bar. See Figure 5.3.1-2

We create a Stick version of the divisor, which in this case is 8, and placing the stick to the left of our divisor will help you envision the approach. In traditional division, you would "guess" how many times the dividend can go into the divisor or the divisor times what dividend is closest to the dividend. Remember, using the Sticks for the divisor eliminates any "guessing." Just look down the rows of the 8 Stick to find a number that is equal to or just smaller than the dividend.

So, let's look at the place values in the dividend (25) that we are going to use. This five-step process was shown in example 1, but it will be shown one more time to clarify where the divisor, dividend, quotient, and remainder come into place.

1. Is the 1st digit of the dividend 2 larger than the divisor (8) which is the first row on the stick?
2. If it is smaller, which it is, then we grab the 2nd digit of the dividend and check it for size.
3. If the new 2-digit dividend is too small, we continue to add a new digit, until it is not.

Now that we have the right number of digits to work with.

4. Now look down the stick, rows 1 through 9 to find a number that is equal to or just smaller than the dividend. In this case, we have 24 on the stick at row 3 just smaller than our dividend of 25.

Place the 3 over the last quotient's digit, 5. Place the answer, 8 x3 = 24, from the stick under the dividend of 25.

5. Then subtract. This time we have a remainder of 1, which means we have one value greater than a perfect multiplication match. As we planned.

So, we can repeat this process by adding a decimal 0 value to the 1 or accepting the remainder of 1. The answer is 3 with a remainder of 1.

Example 3: 54 ÷6.

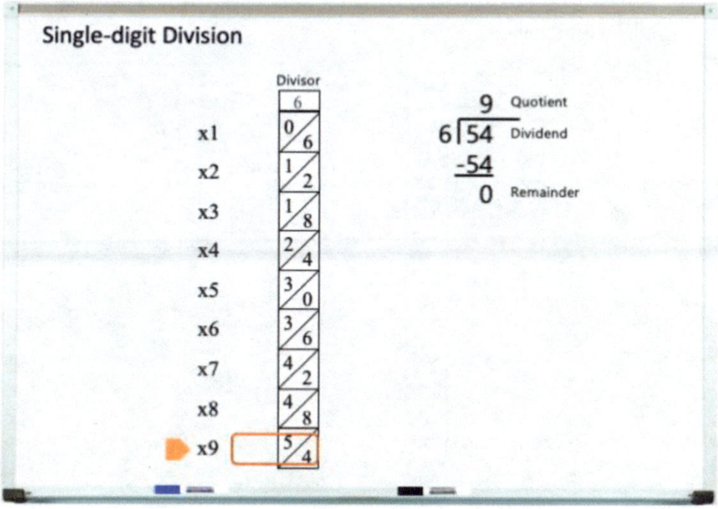

Figure 5.3.1-3: Whiteboard Division: 54 ÷6.

Let's continue with one more problem 6 is the divisor and 54 is the dividend. So, again, we will select the Stick number 6, see Figure 5.3.1-3, to do our problem. Remember to use the divisor in selecting our Tally-x sticks to eliminate any "guessing." Just look down the rows of the Tally-x stick to find a number that is equal to or just smaller than the dividend.

However, before we do that we need to set which place values in the dividend, we are going to use. This was shown in Pattern 11 for division, specifically Chapter 2.3.11.3. We have a 2-digit dividend and a 1-digit divisor.

- Is the 1st digit of the dividend (5) larger than divisor (6) which is the first row on the stick?
- If it is smaller, which it is, then we grab the 2nd digit of the dividend and check it for size.

Now that we have the right number of digits to work with, we look down the divisor stick, rows 1 through 9 to find a number that is equal to or just smaller than the dividend. In this case, we have 54 on the stick in row 9, which is our dividend.

Place the 9 over the last dividend's digit, 4. Place the answer, 6 x9 = 54, from the stick under the dividend of 54. Now subtract. This time we have no remainder, which means we have a perfect quotient match.

The answer is 9.

Example 4: 225 ÷5.

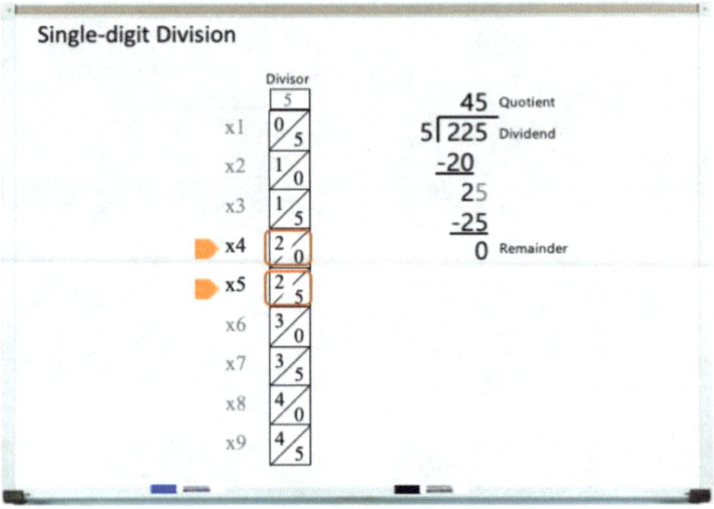

Figure 5.3.1-4: Whiteboard Division: 225 ÷5.

We have done two-digit dividends; so, let's try a three-digit dividend value. This problem will be to divide 5 into 225, see Figure 5.3.1-4. We will follow the same procedures that we did above.

With the right number of digits to work with. We look down the stick, rows 1 through 9 to find a number that is equal to or just smaller than the dividend of 22. In this case, we have 20 on the stick in row 4, which is our dividend.

Place the 4 over the last dividend's digit, 2. Place the answer, 5x4 = 20, from the stick under the dividend of 22. Now subtract, and this time we have a remainder of 2, which means we have a dividend that is greater than a perfect multiplication match. So, we can repeat the process, but this time with a new dividend. This new dividend consists of the remainder plus the next dividend value, which in this case is a 5 and our new dividend is 25.

We repeat steps 1 to 6, but with a dividend of 25, by looking down the stick, rows 1 through 9, to find a number that is equal to or just smaller than the dividend. In this case, we have 25 on the stick at row 5, which is our new quotient on top.

Place the 5 over the last dividends digit, 5. Place the answer, 5 x5 = 25, from the divisor stick under the dividend of 25. Now subtract and this time we have no remainder, which means we have a perfect dividend match.

The answer is 45.

Example 5: 1008 ÷7.

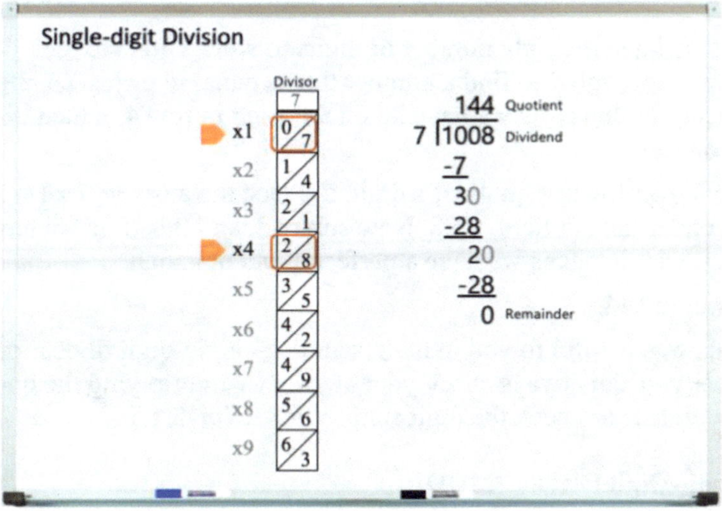

Figure 5.3.1-5: Whiteboard Division: 1008 ÷7.

Now; let's try a four-digit value. This problem will be to divide 7 into 1008, see Figure 5.3.1-5. We will follow the same procedures from above.

With the right number of digits to work with, we look down the divisor stick, rows 1 through 9 to find a number that is equal to or just smaller than the dividend of 10. In this case, we have 7 on the stick at row 1, which is our quotient.

Place the 1 over the last dividend's digit, 0. Place the answer, 7 x1 = 7, from the stick under the dividend of 10. Now subtract, and this time we have a remainder of 3, which means we have a dividend that is greater than a perfect multiplication match. So, we repeat the procedure, but this time with a new dividend. This consists of the remainder 3, plus the next dividend value 0, and our new dividend is 30.

We repeat the procedure steps but with a dividend of 30. So, we look down the stick again in rows 1 through 9 to find a number that is equal to or just smaller than the dividend. In this case, we have 28 on the stick in row 4, which is our new quotient on top.

Place the 4 over the last dividend's digit, 8. Place the answer, 7 x4 = 28, from the stick under the dividend of 30. Now subtract and again this time we have a remainder of 2, which means we have a dividend that is greater than a perfect multiplication match. So, we repeat the procedure again, but this time with a new dividend. This new dividend consists of the remainder plus the next dividend value, which in this case is an 8 and so our new dividend is 28.

We repeat the procedure for the quotient's 1s place value but with a dividend of 28.

Now that we have the right number of digits to work with, we look down the stick, rows 1 through 9 to find a number that is equal to or just smaller than the dividend. In this case, we have 28 on the stick in row 4, which is our new quotient on top.

Place the 4 over the last quotient's digit, 5. Place the answer, 7 x4 = 28, from the stick under the quotient of 28. Now subtract, and this time we have no remainder, which means we have a perfect quotient match.

The answer is 144.

I hope this was helpful to you in understanding single-digit divisor problems. Remember you can always check your answer by multiplying the quotient times the divisor or check the digit sum as in Section 3.1.14.

5.3.2: Multi-Digit Divisor (MDD)

In Section 5.3.1, we showed you how easy it is to do single-digit divisor problems using the Tally-x sticks. In this lesson, we are going to do double-digit division problems using long division. So, let's get out the Tally-x sticks and follow along using your Tally-x sticks.

To review the division place value replacement procedure.

If the 1st digit of the dividend is larger than the divisor, then check the rows on the stick for an answer. If it is smaller, then we grab the 2nd digit of the dividend and check it for size. So, if the new 2-digit quotient is smaller, we continue to add a new dividend digit till it is not.

Example 1: 54 ÷18'

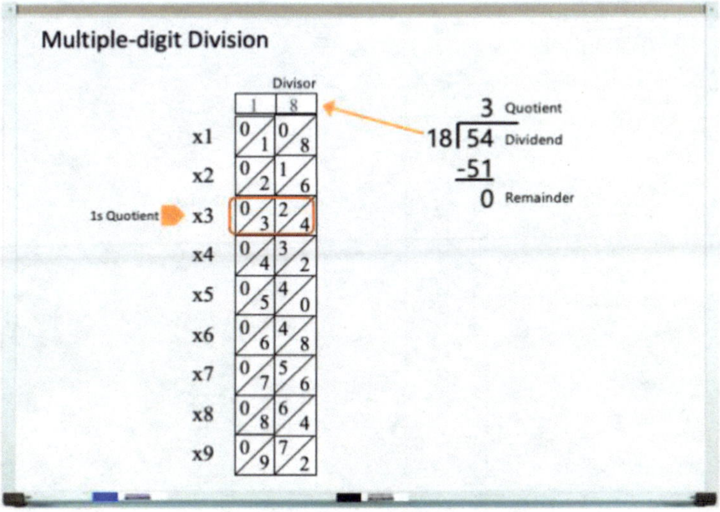

Figure 5.3.2-1: Whiteboard Division: 54 ÷18.

Let's start with our first problem, eighteen into 54. Where eighteen is the divisor, and 54 is the dividend. Here we select the Tally-x stick number, see Figure 5.3.2-1, to do our problem, which is to use the one and eight sticks to form 18.

Now that we have the right number of digits to work with, we look down the stick, rows 1 through 9 to find a number that is equal to or just smaller than the dividend of 54. In this case, we have 54 on the stick at row 3, which is our quotient.

Place the 3 over the last quotient's digit, 4. Place the answer, 18 x3 = 54, from the stick under the quotient of 54. Now subtract, and this time we have no remainder, which means we have a perfect quotient match.

The answer is 3 with no remainder.

Example 2: 53 ÷18.

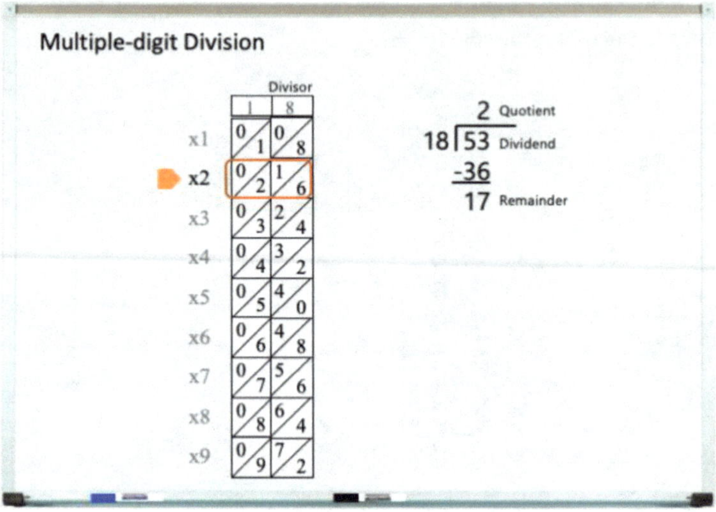

Figure 5.3.2-2: Whiteboard Division: 53 ÷18.

Let's make the previous problem and non-perfect solution, or one with a remainder. Let's use eighteen (18) as the divisor and 53 as the dividend, see Figure 5.3.2-2. So, just as before we will select the Tally-x stick number to do our problem, which is to use the one and eight sticks to form 18.

Now that we have the right number of digits to work with, we look down the stick, rows 1 through 9 to find a number that is equal to or just smaller than the dividend. In this case, we have 36 on the stick at row 2, which is our new quotient on top.

Place the 2 over the last quotient's digit, 3.

Place the answer, 18 x2 = 36, from the stick under the quotient of 53. Now subtract, and this time we have a remainder of 17, which means we have a quotient that is greater than a perfect multiplication match. So, we repeat the procedure again, but this time with a new quotient. This new quotient consists of the remainder plus the next quotient value, which in this case is a decimal 80 and our new quotient is 28.

We can repeat the procedure with the new quotient of 17 and 0 or 170. In this case, we elect to show the answer with the current remainder.

The answer is 2 with a remainder of 17.

Example 3: 174 ÷29.

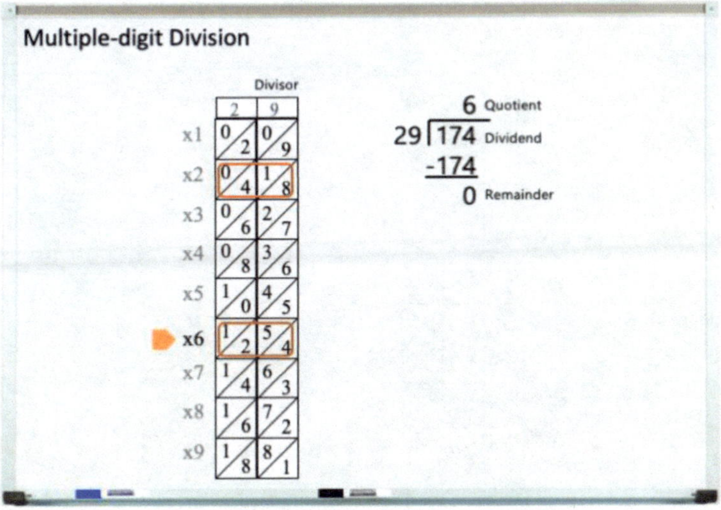

Figure 5.3.2-3: Whiteboard Division: 174 ÷29.

In this problem, our divisor is 29 and the dividend is 174, see Figure 5.3.2-3. Here we select the Tally-x stick number to do our problem, which is the divisor of the two and nine sticks to form 29.

Now that we have the right division components to work with, we look down the stick, rows 1 through 9 to find a number that is equal to or just smaller than the dividend of 174. In this case, we have 174 on the stick at row 6, which is our quotient.

Place the 6 over the last quotient's digit (4) in the long division example. This is done by placing 6 as the quotient and the answer, 29 x6 = 174, from the stick under the dividend of 174. Now subtract, and this time we have no remainder, which means we have a perfect quotient match.

The answer is 6 with no remainder.

Example 4: 1125 ÷125.

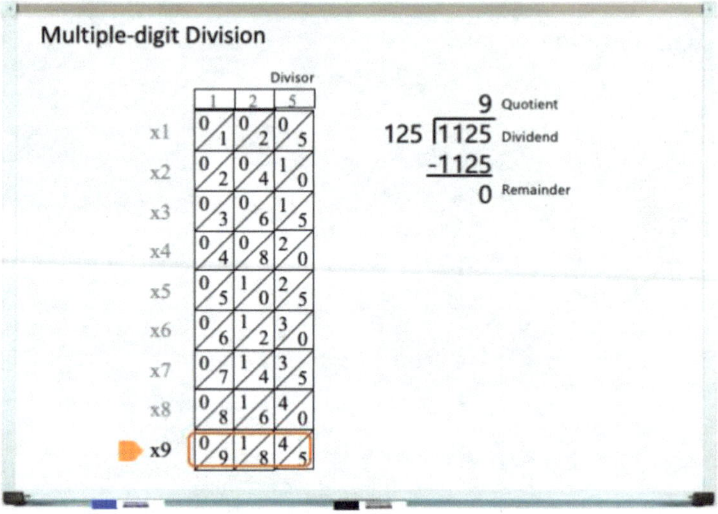

Figure 5.3.2-4: Whiteboard Division: 1125 ÷125.

We have done two- and three-digit dividends; let's try a four-digit value.

In this problem, our divisor is 125, and the dividend is 1125, see Figure 5.3.2-4. Here we select the Tally-x stick number to do our problem, which is the divisor of the one, two, and five sticks to form 125.

Is the 1st digit of the quotient (1) larger than the divisor (125), the first row on the stick?

If it is smaller, which it is, then we grab the 2nd digit of the quotient and check it for size. If the new 2-digit quotient is smaller, we continue to add a new digit, until it is not.

Now that we have the right number of digits to work with, we look down the stick, rows 1 through 9 to find a number that is equal to or just smaller than the dividend of 1125. In this case, we have 1125 on the stick in row 9, which is our quotient.

Place the 9 over the last quotient's digit, 54.

Place the answer, 125 x9 = 1125, from the stick under the quotient of 1125. Now subtract, and again we have no remainder, which means we have a perfect quotient match.

The answer is 9.

Example 5: 72448 ÷342.

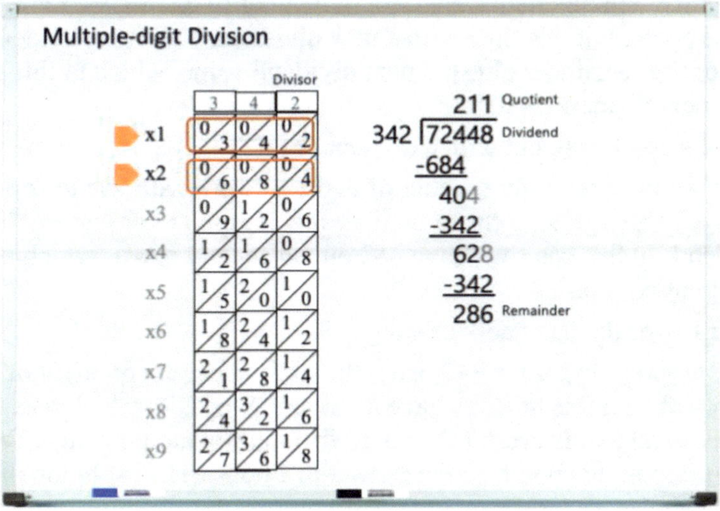

Figure 5.3.2-5: Whiteboard Division: 72448 ÷342.

Let's tackle a five-digit dividend value.

In this problem, our divisor is 342, and the dividend is 72448, as shown in Figure 5.3.2-5. To solve this, we use the Tally-x stick with the three, four, and two sticks representing the divisor 342.

After setting up with the correct number of digits, we proceed by looking down the stick from rows 1 through 9 to find a number that is equal to or just smaller than the dividend 72448. In this case, we find 684 on the stick at row 2, which becomes our initial quotient.

Place the digit 2 over the last digit of the quotient, 4. Next, write down the product 342 x 2 = 684 from the stick under the dividend 72448. After performing the subtraction, we find a remainder of 40, indicating that our quotient exceeds a perfect multiplication match. Therefore, we repeat the process using the new quotient. This consists of the remainder combined with the following quotient value, resulting in 404 as our new quotient.

We then repeat steps 1 through 6 with the quotient of 404. Having the correct number of digits to work with, we again examine the stick, looking from rows 1 through 9 to find a number that is equal to or just smaller than the dividend. In this case, we find 342 on the stick at row 1, establishing this as our new quotient.

Place the digit 1 over the last digit of the quotient, 4.

Place the answer, 342 x 1 = 342, from the stick under the quotient of 404. Now subtract, and this time we have a remainder of 62. So, we repeat the procedure again, but this time with a new dividend. This new dividend consists of the remainder plus the next dividend value, which in this case is 8, so our new dividend is 628.

We repeat steps 1 to 6, but with a dividend of 628.

Now that we have the right number of digits to work with, we look down the stick, rows 1 through 9, to find a number that is equal to or just smaller than the dividend. In this case, we have 342 on the stick at row 1, which is our new quotient on top.

Place the 1 over the last quotient's digit, 8.

Place the answer, 342 x 1 = 342, from the stick under the quotient of 628. Now subtract, and this time we have a remainder of 286, which means we have a dividend that is greater than a perfect multiplication match. We choose to stop at this point, so the answer is a non-perfect solution of 211 with a remainder of 286.

The answer is 211 with a remainder of 286.

I hope this has been helpful to you in understanding multiple-digit divisor problems. Try this on your own by using examples of multiple-digit divisors or make them up on your own. Remember you can always check your answer by multiplying the quotient times the divisor.

5.4: Division Worksheet.

5.4.1: Division Worksheet 1

The following problem sets are single-digit divisor values into multi-digit dividends. Answers are provided in parentheses (…). There is no remainder so "R" is not shown.

1. 1,083 ÷ 3 = _______	(361)	26. 3,780 ÷ 3 = _______	(1,260)
2. 5,067 ÷ 9 = _______	(563)	27. 8,975 ÷ 5 = _______	(1,795)
3. 1,664 ÷ 2 = _______	(832)	28. 1,224 ÷ 2 = _______	(612)
4. 5,931 ÷ 9 = _______	(659)	29. 7,660 ÷ 5 = _______	(1,532)
5. 7,664 ÷ 8 = _______	(958)	30. 1,448 ÷ 4 = _______	(362)
6. 8,045 ÷ 5 = _______	(1,609)	31. 4,134 ÷ 6 = _______	(689)
7. 9,779 ÷ 7 = _______	(1,397)	32. 502 ÷ 2 = _______	(251)
8. 5,212 ÷ 4 = _______	(1,303)	33. 7,635 ÷ 5 = _______	(1,527)
9. 448 ÷ 8 = _______	(56)	34. 3,290 ÷ 7 = _______	(470)
10. 468 ÷ 6 = _______	(78)	35. 198 ÷ 6 = _______	(33)
11. 4,854 ÷ 6 = _______	(809)	36. 5,250 ÷ 6 = _______	(875)
12. 1,816 ÷ 4 = _______	(454)	37. 5,340 ÷ 3 = _______	(1,780)
13. 5,184 ÷ 6 = _______	(864)	38. 912 ÷ 8 = _______	(114)
14. 4,953 ÷ 3 = _______	(1,651)	39. 8,232 ÷ 2 = _______	(4,116)
15. 96 ÷ 4 = _______	(24)	40. 8,911 ÷ 7 = _______	(1,273)
16. 1,190 ÷ 5 = _______	(238)	41. 81 ÷ 9 = _______	(9)
17. 3,984 ÷ 8 = _______	(498)	42. 6,748 ÷ 4 = _______	(1,687)
18. 2,030 ÷ 2 = _______	(1,015)	43. 2,280 ÷ 8 = _______	(285)
19. 6,150 ÷ 3 = _______	(2,050)	44. 7,380 ÷ 6 = _______	(1,230)
20. 5,450 ÷ 5 = _______	(1,090)	45. 1,287 ÷ 3 = _______	(429)
21. 7,392 ÷ 7 = _______	(1,056)	46. 5,789 ÷ 7 = _______	(827)
22. 2,988 ÷ 9 = _______	(332)	47. 2,916 ÷ 2 = _______	(1,458)
23. 6,624 ÷ 6 = _______	(1,104)	48. 3,296 ÷ 8 = _______	(412)
24. 2,940 ÷ 4 = _______	(735)	49. 8,824 ÷ 2 = _______	(4,412)
25. 3,042 ÷ 9 = _______	(338)	50. 9,306 ÷ 9 = _______	(1,034)

5.4.2: Division Worksheet 2

The following problem sets are single-digit divisor values into multi-digit dividends. Answers are provided in parentheses (…). There is no remainder so "R" is not shown.

1. 1,693 ÷ 3 = _______ (564)
2. 8,055 ÷ 9 = _______ (895)
3. 8,480 ÷ 2 = _______ (4,240)
4. 4,032 ÷ 9 = _______ (448)
5. 4,512 ÷ 8 = _______ (564)
6. 5,015 ÷ 5 = _______ (1,003)
7. 7,152 ÷ 7 = _______ (1,022)
8. 2,936 ÷ 4 = _______ (732)
9. 5,664 ÷ 8 = _______ (708)
10. 5,010 ÷ 6 = _______ (835)
11. 1,060 ÷ 2 = _______ (530)
12. 8,104 ÷ 4 = _______ (2,026)
13. 1,404 ÷ 6 = _______ (234)
14. 82,893 ÷ 3 = _______ (27,631)
15. 916 ÷ 4 = _______ (229)
16. 2,905 ÷ 5 = _______ (581)
17. 6,416÷ 8 = _______ (802)
18. 1,330 ÷ 2 = _______ (665)
19. 5,568 ÷ 3 = _______ (1,856)
20. 5,395 ÷ 5 = _______ (1,079)
21. 6,020 ÷ 7 = _______ (860)
22. 6,579 ÷ 9 = _______ (731)
23. 4,224 ÷ 6 = _______ (704)
24. 2,444 ÷ 4 = _______ (611)
25. 891 ÷ 9 = _______ (99)
26. 4,572÷ 3 = _______ (1,524)
27. 7,800 ÷ 5 = _______ (1,560)
28. 6,699 ÷ 7 = _______ (957)
29. 7,788 ÷ 2 = _______ (3,894)
30. 9,916 ÷ 4 = _______ (2,479)
31. 6,786 ÷ 6 = _______ (1,131)
32. 1,697 ÷ 2 = _______ (849)
33. 4,175 ÷ 5 = _______ (835)
34. 2,023 ÷ 7 = _______ (289)
35. 3,024 ÷ 8 = _______ (378)
36. 6,600 ÷ 6 = _______ (1,100)
37. 3,240 ÷ 2 = _______ (1,620)
38. 5,910 ÷ 3 = _______ (1,970)
39. 3,032 ÷ 8 = _______ (379)
40. 5,972 ÷ 2 = _______ (2,986)
41. 4,938 ÷ 6 = _______ (823)
42. 9,369 ÷ 9 = _______ (1,041)
43. 4,380 ÷ 4 = _______ (1,095)
44. 1,218 ÷ 2 = _______ (609)
45. 1,218 ÷ 5 = _______ (1,123)
46. 5,615 ÷ 6 = _______ (1,226)
47. 9,798 ÷ 9 = _______ (3,266)
48. 3,815 ÷ 7 = _______ (545)
49. 2,758 ÷ 2 = _______ (1,379)
50. 1,620 ÷ 9 = _______ (180)

5.4.3: Division Worksheet 3

The following problem sets are double-digit divisor values into multi-digit dividends. Answers are provided in parentheses (…). followed by any remainder "R".

1. 7122 ÷ 15 = ______ (474) R 12
2. 7652 ÷ 72 = ______ (106) R 20
3. 4363 ÷ 38 = ______ (114) R 31
4. 9728 ÷ 81 = ______ (120) R 8
5. 2374 ÷ 49 = ______ (48) R 22
6. 5406 ÷ 43 = ______ (125) R 31
7. 8523 ÷ 29 = ______ (293) R 26
8. 3103 ÷ 44 = ______ (70) R 23
9. 8202 ÷ 88 = ______ (93) R 18
10. 647 ÷ 28 = ______ (23) R 3
11. 8921 ÷ 17 = ______ (524) R 13
12. 2233 ÷ 38 = ______ (58) R 29
13. 6471 ÷ 65 = ______ (99) R 36
14. 7715 ÷ 42 = ______ (183) R 29
15. 6349 ÷ 85 = ______ (74) R 59
16. 1196 ÷ 24 = ______ (49) R 20
17. 2977 ÷ 73 = ______ (40) R 57
18. 9697 ÷ 74 = ______ (131) R 3
19. 7290 ÷ 76 = ______ (95) R 70
20. 3316 ÷ 49 = ______ (67) R 22
21. 1215 ÷ 92 = ______ (13) R 19
22. 8896 ÷ 97 = ______ (91) R 69
23. 5294 ÷ 25 = ______ (211) R 19
24. 6888 ÷ 71 = ______ (97) R 1
25. 7148 ÷ 74 = ______ (96) R 44
26. 9017 ÷ 14 = ______ (644) R 1
27. 7500 ÷ 53 = ______ (141) R 27
28. 3094 ÷ 74 = ______ (41) R 60
29. 8385 ÷ 33 = ______ (254) R 3
30. 833 ÷ 39 = ______ (21) R 14
31. 4575 ÷ 6 = ______ (762) R 3
32. 5209 ÷ 29 = ______ (179) R 18
33. 6240 ÷ 15 = ______ (416) R 0
34. 81 ÷ 68 = ______ (1) R 13
35. 4244 ÷ 85 = ______ (49) R 79
36. 8919 ÷ 35 = ______ (254) R 29
37. 6226 ÷ 91 = ______ (68) R 38
38. 1380 ÷ 47 = ______ (29) R 17
39. 3542 ÷ 57 = ______ (62) R 8
40. 7946 ÷ 70 = ______ (113) R 36
41. 475 ÷ 63 = ______ (7) R 34
42. 1491 ÷ 74 = ______ (20) R 11
43. 439 ÷ 46 = ______ (9) R 25
44. 6339 ÷ 68 = ______ (93) R 15
45. 8906 ÷ 13 = ______ (685) R 1
46. 6575 ÷ 8 = ______ (821) R 7
47. 6310 ÷ 91 = ______ (69) R 31
48. 6659 ÷ 53 = ______ (125) R 34
49. 7184 ÷ 19 = ______ (378) R 2
50. 6050 ÷ 97 = ______ (62) R 136

5.4.4: Division Worksheet 4

The following problem sets are double-digit divisor values into multi-digit dividends. Answers are provided in parentheses (…) followed by any remainder “R”.

1. 7484 ÷ 62 = ______	(120) R 44	26. 406 ÷ 26 = ______	(15) R 16
2. 6459 ÷ 91 = ______	(70) R 89	27. 6655 ÷ 31 = ______	(214) R 21
3. 5633 ÷ 9 = ______	(625) R 8	28. 9198 ÷ 69 = ______	(133) R 21
4. 3080 ÷ 79 = ______	(38) R 78	29. 7717 ÷ 86 = ______	(89) R 63
5. 9916 ÷ 27 = ______	(367) R 7	30. 1056 ÷ 27 = ______	(39) R 3
6. 6015 ÷ 40 = ______	(150) R 15	31. 8213 ÷ 10 = ______	(821) R 3
7. 9813 ÷ 4 = ______	(2453) R 1	32. 1482 ÷ 15 = ______	(98) R 12
8. 3703 ÷ 6 = ______	(617) R 1	33. 3544 ÷ 83 = ______	(42) R 58
9. 8330 ÷ 39 = ______	(213) R 23	34. 7202 ÷ 12 = ______	(600) R 2
10. 9983 ÷ 59 = ______	(169) R 12	35. 1805 ÷ 17 = ______	(106) R 3
11. 6912 ÷ 64 = ______	(108) R 0	36. 6343 ÷ 31 = ______	(204) R 19
12. 6470 ÷ 30 = ______	(215) R 20	37. 205 ÷ 29 = ______	(7) R 2
13. 309 ÷ 49 = ______	(6) R 15	38. 3768 ÷ 56 = ______	(67) R 16
14. 1593 ÷ 40 = ______	(39) R 33	39. 5076 ÷ 78 = ______	(65) R 6
15. 7046 ÷ 36 = ______	(195) R 26	40. 9059 ÷ 48 = ______	(188) R 35
16. 9205 ÷ 95 = ______	(96) R 85	41. 838 ÷ 94 = ______	(8) R 86
17. 4552 ÷ 63 = ______	(72) R 16	42. 2335 ÷ 35 = ______	(66) R 25
18. 2807 ÷ 35 = ______	(80) R 7	43. 5910 ÷ 23 = ______	(256) R 22
19. 1533 ÷ 31 = ______	(49) R 14	44. 971 ÷ 49 = ______	(19) R 40
20. 5105 ÷ 98 = ______	(52) R 9	45. 9870 ÷ 56 = ______	(176) R 14
21. 5340 ÷ 42 = ______	(127) R 6	46. 6304 ÷ 43 = ______	(146) R 26
22. 8723 ÷ 31 = ______	(281) R 12	47. 1661 ÷ 79 = ______	(21) R 2
23. 6551 ÷ 68 = ______	(96) R 23	48. 8620 ÷ 63 = ______	(136) R 52
24. 6644 ÷ 86 = ______	(77) R 22	49. 33781 ÷ 71 = ______	(475) R 56
25. 6859 ÷ 46 = ______	(149) R 5	50. 13654 ÷ 41 = ______	(240) R 12

Chapter 6: Decimal Multiplication and Division

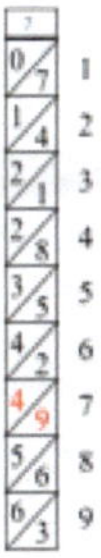

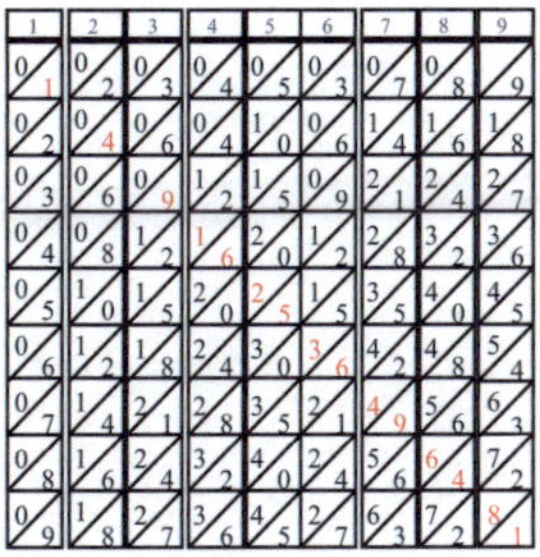

6.5 → 1 Decimal Place

x 7 → 0 Decimal Place

35

+42

455

45.5 ← 1 Decimal Place

9.7 → 1 Decimal Place

x 4 → 0 Decimal Place

28

36

388

38.8 ← 1 Decimal Place

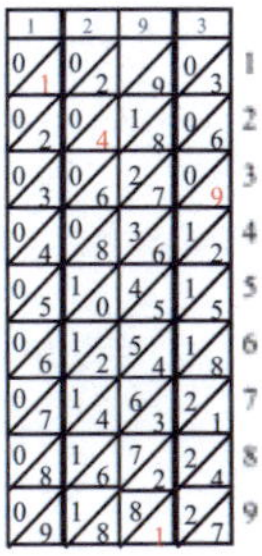

1.293 → 3 Decimal Place

x 2.7 → 1 Decimal Place

9051

2586

34911

3.4911 ← 4 Decimal Place

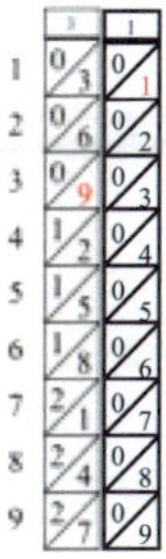

0.31 ⟌1083. 31.⟌108300.

Shift to Balance

3493.5

31.⟌108300.

93 ← x3

153

124 ← x4

290

279 ← x9

110

93 ← x3

170

155 ← x5

15

49.8⟌971. 498.⟌9710.

Shift to Balance

19.49

498.⟌9710.

498 ← x1

4730

4482 ← x9

2480

1992 ← x4

4880

4482 ← x9

398

6.1: Decimal Math Introduction.

6.1.1 Fractions.

In this chapter we will address the idea that numbers are not always whole (integer) values: 0, 1, 2, 3, 4, 5, 6, 7, 8, 9. This was, or should have been, taught in adding and subtraction. A decimal is the partial value between two whole numbers or a fraction. For example, halfway between 2 and 3 is 2½, 5 and 6 is 5½, 8 and 9 is 8½. Note that the whole numbers are 1 value apart and consist of the lower whole number and the partial value between the two whole numbers.

Fractions are expressed as a numerator (on top) and a denominator (on bottom). In the half fraction above, the denominator is 2, and the numerator is 1, which shows '1 divided by 2'. This implies the fraction is halfway between the two whole numbers.

One could think of this as dividing a whole pie into equal pieces of a pie over and over, as in Figure 6.1.1-1 below.

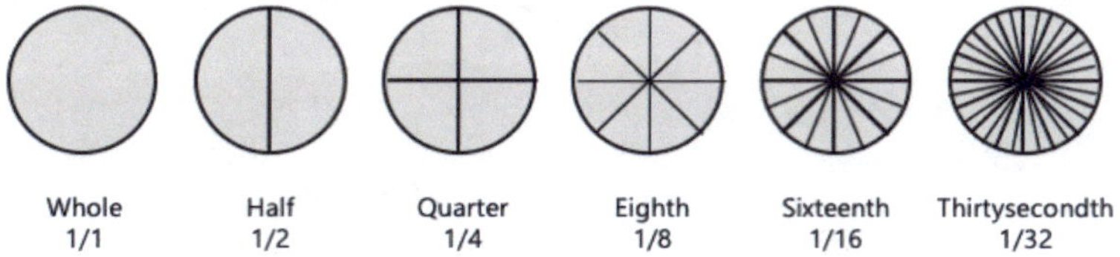

Figure 6.1.1-1 Divisions of a Pie\e.

6.1.2 Decimals.

Decimals are similar to fractions, but they differ in that decimals can represent all the values between whole numbers, like the slices of a pie cut into very small pieces. For instance, halfway between 2 and 3 is expressed as 2.5, between 5 and 6 as 5.5, and between 8 and 9 as 8.5. This concept is akin to fractions where 0.5 equals half (½) between 0 and 1. However, decimals can also represent values that are not fractions, such as 2.11, 2.123456, or numbers with infinitely many decimal places.

One specific constant, known as "Pi" (π), is particularly important in geometry. Pi represents the unchanging relationship between a circle's diameter and its circumference—the distance around the circle. This value is infinitely long and never repeats, making it a challenging but essential concept for students to grasp. Figure 6.1.2-1 shows the first 50 digits of Pi, but even remembering 10 of them can be difficult because Pi has no discernible pattern and is non-repeating.

3.1415926535 8979323846 2643383279 5028841971 6939937510

Figure 6.1.2-1 Value of Pi to 50 decimals.

6.1.3 Decimals and Fractions.

In this chapter, we will address the idea that numbers are not always whole (integer) values: 0, 1, 2, 3, 4, 5, 6, 7, 8, 9. But, can be a value between two whole numbers, or a fraction. For example, halfway between 2 and 3 is 2.5, 5 and 6 is 5.5, 8 and 9 is 8.5. Note that these whole numbers are 1 value apart. The idea is similar to a fraction value where 0.5 is the same as ½ fraction between 0 and 1.

Fraction		Division		Division		Decimal		Decimal
1/2	*or*	1 ÷2	*or*	$2\overline{)1.0}$ = 0.5	=	0.5	*or*	0.50000
1/4	*or*	1 ÷4	*or*	$4\overline{)1.00}$ = 0.25	=	0.25	*or*	0.25000
1/8	*or*	1 ÷8	*or*	$8\overline{)1.000}$ = 0.125	=	0.125	*or*	0.12500
1/16	*or*	1 ÷16	*or*	$16\overline{)1.0000}$ = 0.0625	=	0.0625	*or*	0.06250
1/32	*or*	1 ÷32	*or*	$32\overline{)1.00000}$ = 0.03125	=	0.03125	*or*	0.03125

Table 6.1.3-1 Decimals and Fractions.

In Table 6.1.3-1 you can see that a fraction can be expressed as a decimal value by dividing the lower value into the upper value. I.e., 1 ÷2 is 0.5 and 1 ÷16 is 0.06250. In the table, each row is an equivalent across the row value: ½ or 1 ÷2 or 0.5 is the same as the division equation.

This section would be best taught by understanding the first two columns: the fraction and its equivalent decimal in Chapter 5. Then, followed by the third column of division, which is found in chapter 6.1.4.2 below.

6.1.4 Traditional Decimal Multiplication.

Traditional decimal multiplication or long multiplication involves 1) carrying super scripts, 2) multiplication, 3) ignoring the decimals all together, and 4) finally with all the multiplication and additions completed, counting the number of decimals in the multiplicand and multiplier (if any), 5) Then placing the decimal by counting from the right into the answer.

This method works, but it adds a few complexities to the problem:

- Memorizing the multiplication table, 0 to 9.
- Learning to carry the tens value above the next multiplicand, and erasing it with each new multiplier.

```
 6.5   → 1 Decimal Place (DP)
 x 7   → 0 DP
 35      +
  ↓      ↓
 3.5   ← 1 DP
```

Figure 6.1.4-1 Traditional Decimal Multiplication.

This approach resolves the multiplication complexities by reversing the order and counting the decimal value first. It also removes the complexity of carrying and erasing superscripts. This way you are dealing with a whole number just like in Chapter 4. Only in this case, do you add the counted decimals back to the answer at the end.

6.1.5 Traditional Decimal Division.

Traditional decimal division or long division involves 1) carrying super scripts, 2) division, 3) ignoring the decimals together, 4) counting the number of decimals in the multiplicand and multiplier (if any), and 5) placing the decimal by counting from the right into the answer.

```
                          19
4.9 ⌈971    ⇒    49 ⌈9710.
```

Count *left divisor* decimal places (DR) on left -> shift *right dividend* same DR count

```
       193 r 8
49  ⌈9710
     49
     481
     441
      400
      392
        8
```

Figure 6.1.5-1 Traditional Decimal Division.

This method works, but it adds a few complexities to the problem, see the points in section 6.1.4 above.

This approach resolves the division complexities by reversing the order and counting the decimal value first. It also eliminates the complexity of carrying and erasing exponents. This way, you are dealing with a whole number similar to Chapter 5. Here, you simply add the counted decimals back to the answer at the end.

6.2: Tally-x Decimal Multiplication.

6.2.1: Multiplication Place Value Counting.

In Figure 6.1.4-1, you see that the first step is to access the use of a decimal marker between the integer value and the decimal value for both the multiplicand and multiplier. This is not necessary if either or both of the multiplicand and multiplier are integer values.

If either of the multiplicand and multiplier are decimal values, count the total number of values to the right of the decimal for each. As seen in 6.4.1-1, that is the action at the top of the figure. This results in the total number of decimal places (DP) of the problem.

Now remove the decimal marker, or ignore them, for both the multiplicand and multiplier. Write this down somewhere or just remember the DP value. See Figure 6.2.2-1 below for an example. Now treat the multiplication as an integer, as in Section 4: Multiplication.

When you get the result or product, you count the DP from the right of the product and put a decimal marker into the product. This is seen in the lower-right part of Figure 6.2.2-1.

6.2.2: Tally-x Stick Decimal Multiplication.

Using Figure 6.2.2-1, you can apply the basic Place Value Counting rules from above. The multiplicand becomes the equivalent 7-stick. The integer version of the multiplier is the row number, such as in case 7. Just as described in Section 4: Multiplication.

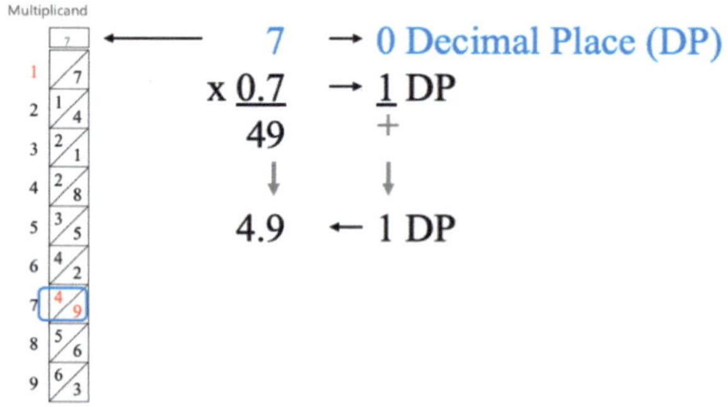

Figure 6.2.2-1 Tally-x Stick Decimal Multiplication: 7 x0 .7.

The answer in the 7 stick is 49. To this, you put back the decimal place value of 1, moving it between the 10s and 1s value, 4.9.

In Figure 6.2.2-2, you apply the basic Place Value Counting rules. The multiplicand becomes the 6s and 8s stick for the 10s and 1s value. The integer version of the multiplier is the row number in this case 5. Just as described in Section 4: Multiplication.

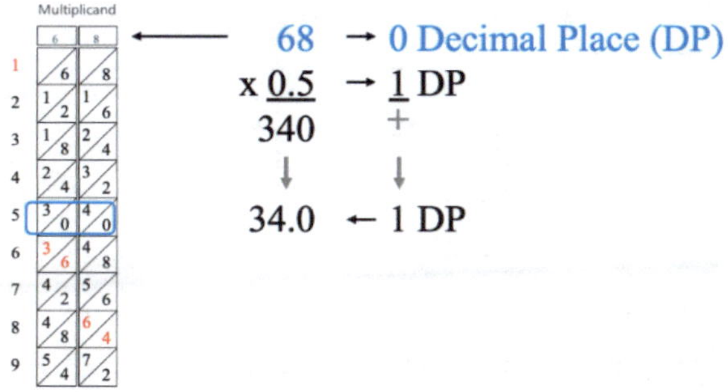

Figure 6.2.2-2 Tally-x Stick Decimal Multiplication: 68 x0.5.

The answer in the 6 8 sticks is 340. To this, you put back the decimal place value of 1, moving it between the 10s and 1s value, 34.0.

In Figure 6.2.2-3, you apply again the basic Place Value Counting rules. The multiplicand becomes the 9s and 7s sticks for the 10s and 1s values. The integer version of the multiplicand is 97, with a DP value of 1. The integer version of the multiplier is 8, without a DP value.

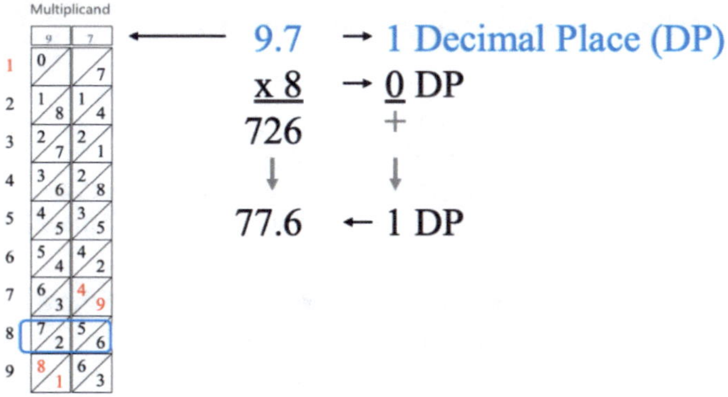

Figure 6.2.2-3 Tally-x Stick Decimal Multiplication: 9.7 x8.

The answer in the 9 7 sticks is 776. To this, you put back the decimal place value of 1, moving it between the 10s and 1s value, 77.6.

In Figure 6.2.2-4, using basic Place Value Counting rules, the multiplicand becomes the 4s, 2s, 2s, and 8s sticks for the 1000s, 100s, 10s, and 1s values. The multiplicand is already an integer, DP =0. The integer version of the multiplier is the row number 57, with a DP =2.

Note: here you don't have multiple rows as you do in traditional multiplication, as the various place values are all shown in the 5th and 7th rows. Thus making it simpler.

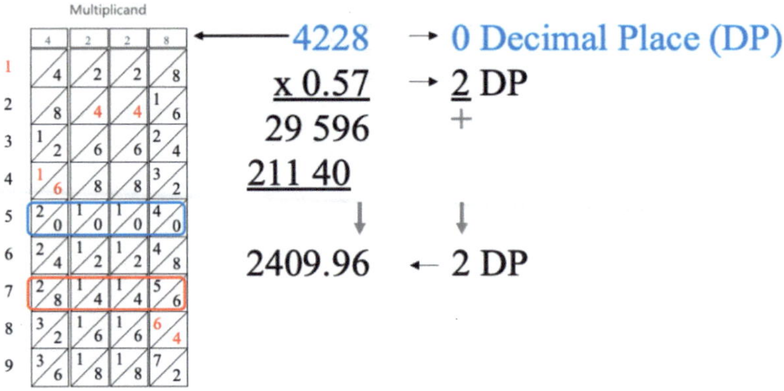

Figure 6.2.2-4 Tally-x Stick Decimal Multiplication: 4228 x0.57.

The answer in the 4 2 2 87 sticks is 240996. To this, you put back the decimal place value of 2, moving it between the 10s and 1s value, 2409.96.

6.2.3: Tally-x Stick Decimal Multiplication - Summary.

The steps are simple:

- Count the number of decimal places right of the decimal marker, for both the multiplicand and multiplier.
- Do the Tally-x multiplication and ignore the decimal markers.
- In the solution, count from the right the number decimal places, and reinsert the decimal marker.

Counting the decimals is like adding all the place values of the multiplier and multiplicand, and putting it back into the solution.

6.3: Tally-x Decimal Division.

6.3.1 Division Place Value Counting.

In Figure 6.1.5-1, you can see that the first step is to consider the use of a decimal point between the integer and decimal parts for both the dividend and divisor. This step is unnecessary if either or both the dividend and divisor are integers.

If either the dividend or divisor is a decimal value, count the total number of zeros to the right of the decimal for each. This action is shown at the top of Figure 6.1.5-1, and it determines the total number of decimal places (DP) in the problem.

Next, remove or ignore the decimal point for both the dividend and divisor. Note down this information or simply remember the DP value.

Now, proceed with the division process as outlined in Section 5: Division. This involves treating the division as if you were adding integers of the dividend and divisor.

Once you have the quotient, count the DP from the right of the solution and place a decimal point back in the correct position. This is illustrated in the lower part of the figure.

6.3.2: Division Tally-x Stick Decimal Division.

In Figure 6.3.2-1, the color squares are blue (1s), red (10s), green (100s), and purple (1000s); to provide some clarity of order. But, is not necessary in practice.

In this problem only 0.4 (divider) has a shifting decima, DP=1, so we save that till after the division is completed. This is similar to the traditional division in Figure 6.2.2-4, but it is simplified by the 4-stick. As you get to understand the patterns, in Section 3: Patterns of the Multiplication Table, this gets easier.

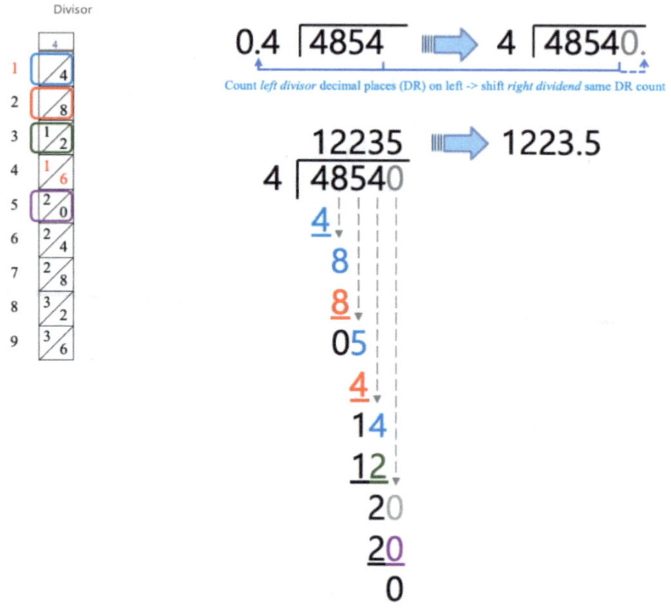

Figure 6.3.2-1 Tally-x Stick Decimal Division: 4854 ÷ 0.4.

In this problem, only 0.9 (dividend) has a shifting decimal, DP=1, so we save that till after the division and then put it back. We could continue the division process to get more and more accuracy, but typically you only go to the same decimal place or one smaller.

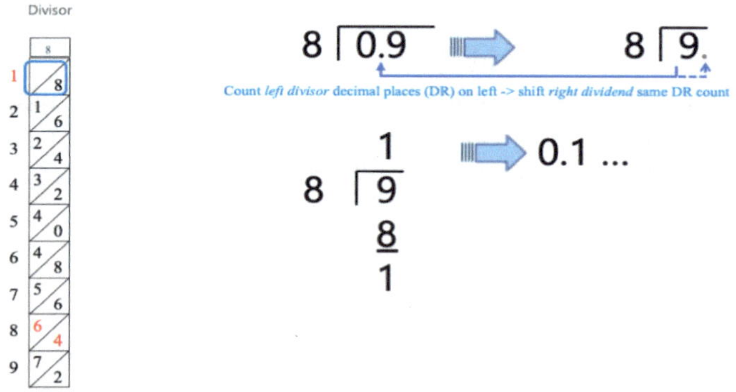

Figure 6.3.2-2 Tally-x Stick Decimal Division: 0.9 ÷ 8.

The answer for 9 ÷8 in integer division is 1 with a remainder of 1. Replacing the DP to the solution gives you 0.1 with a remainder of 1. We could repeat the division process and get a more accurate solution: 1.125.

In this next problem, only 3.6 (divider) has a shifting decimal, DP=1. But, only with a divisor DP can simplify the problem by using the DP value to shift the dividend by adding that many zeros. In this case, the dividend becomes 70460 by adding 1-more zero.

The answer for 70460 ÷36 is an integer division. In this case, we complete the dividend integer repeat four times because that is the number of values in the dividend. The first quotient is 1 and 70-36 is 34, we bring down the next value of the dividend to get 344. The next quotient is therefore 36 x9 or 324 with a remainder of 20, and we repeat this to the last dividend 0 giving 260. This is the 7th row of the divisor or 252. This leaves a remainder of 8.

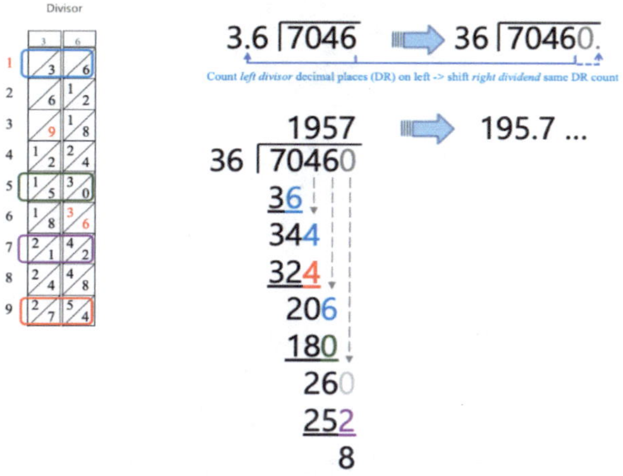

Figure 6.3.2-3 Tally-x Stick Decimal Division: 7046 ÷ 3.6.

We put the DP of 1 back and we get: 195.7 with a remainder of 8.

In the next Figure 6.3.2-4, the color squares are the same, but this is not necessary in practice.

In the next Figure 6.3.2-4, we have a problem is the dividend (4853.54) with two shifting decimals DP=2. But the divisor has none, DP=0. So we have a division of 48535400 ÷36.

We can continue the division process to get more and more accuracy, but typically you only go to the same decimal place or one smaller. Resulting by repeated means a quotient of 1386825 with a remainder of 25.

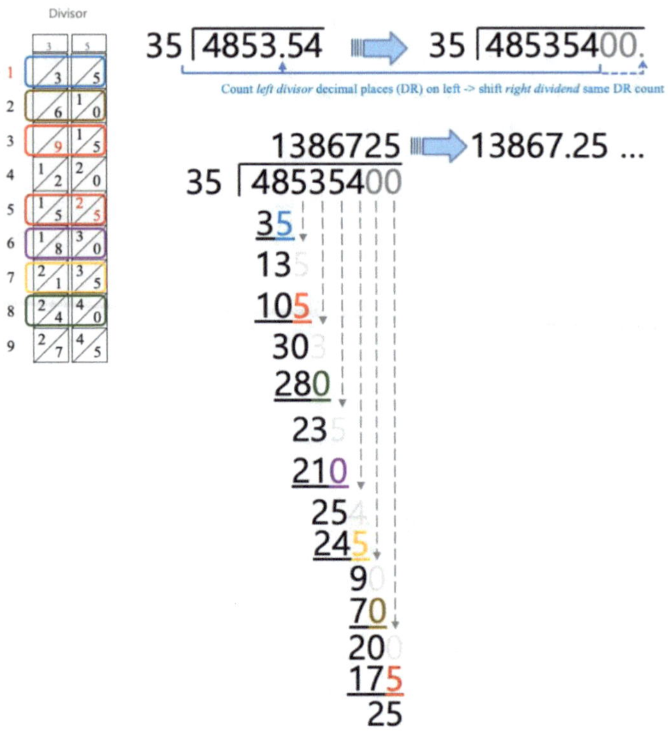

Figure 6.3.2-4 Tally-x Stick Decimal Division: 4853.54 ÷ 35.

Then put the shifting decimals, DP of 2, back to yield 13,8675.25. If the division sequence is continued, the values after the 0.25 would repeat every 6 values as: 7, 1, 4, 2, 8, 5.

6.3.3: Tally-x Stick Decimal Division Summary

The steps are simple and similar to the Tally-x Decimal multiplication approach:

- Count the number of decimal places right of the decimal marker, for both the divider and dividend.
- Do the division process of an integer and ignore the decimal markers.
- In the solution, count from the right the number decimal places, and reinsert the decimal marker.

Chapter 6 Endnote.

6.1 [The universe] cannot be read until we have learned the language and become familiar with the characters in which it is written. *It is written in mathematical language, and the letters are triangles, circles, and other geometrical figures, without which means it is humanly impossible to comprehend a single word,* Citation: *Opere Il Saggiatore* (English: *The Assayer*), p. 171, Roman, 1623 AD. PDF English translation: https://web.stanford.edu/~jsabol/certainty/readings/Galileo-Assayer.pdf

Book Overview.

Let me conclude with the following.

Galileo Galilei was born in 1564 in the same city as Leonardo de Pisa but 394 years later (see Chapter 1). With the advantage of math from Leonardo and his passion for science, Galileo became the father of many things: astronomy, modern physics, scientific methods, and many other sciences.

Galileo's greatest contribution was the math of science. He is quoted to have said in his "Book of Nature".

"It is written in mathematical language, and its characters are triangles, circles, and other geometric figures, without which it is humanly impossible to understand a word; without these, one is wandering in a dark labyrinth."
Reference 6.1

Although patterns are not in Galileo's quote, it is implied, and therefore, as this book has shown, I assert that Patterns are part of the *mathematical language*.

I challenge you to find new patterns in your math studies. As the question, no one else dares to ask, and don't let anyone tell you it was a stupid question. It took me about 40 years to find my answer, "Why memorize the multiplication of 0 to 12?"

You may be the next person to find a new pattern for Chapter 3 or the other five math curriculums (algebra, trigonometry, or calculus). But, don't stop there, you are not limited to math, find a pattern in science that is explained by a new math and a new way of thinking.

6.4: Decimal Multiplication and Division Worksheets

The following Multiplication and Division problems are similar setup as in Section 4: Multiplication and Section 5: Division worksheets, except the answers are in decimal solutions.

6.4.1: Decimal Multiplication Worksheet 1

The following problem sets include multiplicands with double-digit values and dividends with multi-digit values. Answers are provided in parentheses (...).

1. 3 x 0.7 = ______	(2.1)	26. 0.7 x 9 = ______	(6.3)
2. 8 x 0.6 = ______	(4.8)	27. 0.9 x 4 = ______	(3.6)
3. 5 x 0.2 = ______	(1.0)	28. 0.3 x 1 = ______	(0.3)
4. 7 x 0.9 = ______	(6.3)	29. 0.2 x 2 = ______	(0.4)
5. 1 x 0.2 = ______	(0.2)	30. 0.5 x 9 = ______	(4.5)
6. 5 x 0.7 = ______	(3.5)	31. 0.3 x 1 = ______	(0.3)
7. 2 x 0.8 = ______	(1.6)	32. 0.7 x 7 = ______	(4.9)
8. 7 x 0.2 = ______	(1.4)	33. 0.4 x 1 = ______	(0.4)
9. 4 x 0.9 = ______	(3.6)	34. 0.6 x 3 = ______	(1.8)
10. 5 x 0.8 = ______	(4.0)	35. 0.9 x 8 = ______	(7.2)
11. 2 x 0.4 = ______	(0.8)	36. 0.4 x 5 = ______	(2.0)
12. 4 x 0.3 = ______	(1.2)	37. 0.9 x 5 = ______	(4.5)
13. 8 x 0.9 = ______	(7.2)	38. 0.9 x 2 = ______	(1.8)
14. 3 x 0.4 = ______	(1.2)	39. 0.5 x 2 = ______	(1.0)
15. 8 x 0.4 = ______	(3.2)	40. 0.5 x 4 = ______	(2.0)
16. 6 x 0.1 = ______	(0.6)	41. 0.5 x 2 = ______	(1.0)
17. 1 x 0.9 = ______	(0.9)	42. 0.6 x 2 = ______	(1.2)
18. 4 x 0.3 = ______	(1.2)	43. 0.2 x 5 = ______	(1.0)
19. 8 x 0.7 = ______	(5.6)	44. 0.7 x 4 = ______	(2.8)
20. 7 x 0.3 = ______	(2.1)	45. 0.9 x 1 = ______	(0.9)
21. 7 x 0.2 = ______	(1.4)	46. 0.7 x 8 = ______	(5.6)
22. 5 x 0.2 = ______	(1.0)	47. 0.1 x 4 = ______	(0.4)
23. 3 x 0.7 = ______	(2.1)	48. 0.7 x 7 = ______	(4.9)
24. 1 x 0.2 = ______	(0.2)	49. 0.2 x 1 = ______	(0.2)
25. 8 x 0.6 = ______	(4.8)	50. 0.1 x 9 = ______	(0.9)

6.4.2: Decimal Multiplication Worksheet 2

The following problem sets are single-digit multiplier values times a double-digit multiplicand. Answers are provided in parenthesis (…).

1. 99 x 0.1 = ______ (9.9)
2. 58 x 0.9 = ______ (52.2)
3. 30 x 0.9 = ______ (27.0)
4. 22 x 0.5 = ______ (11.0)
5. 44 x 0.4 = ______ (17.6)
6. 50 x 0.4 = ______ (20.0)
7. 91 x 0.9 = ______ (81.9)
8. 47 x 0.2 = ______ (9.4)
9. 86 x 0.9 = ______ (77.4)
10. 60 x 0.6 = ______ (36.0)
11. 55 x 0.2 = ______ (11.0)
12. 34 x 0.2 = ______ (6.8)
13. 45 x 0.4 = ______ (18.0)
14. 51 x 0.1 = ______ (5.1)
15. 86 x 0.7 = ______ (60.2)
16. 47 x 0.3 = ______ (14.1)
17. 63 x 0.3 = ______ (18.9)
18. 43 x 0.8 = ______ (34.4)
19. 62 x 0.1 = ______ (6.2)
20. 16 x 0.6 = ______ (9.6)
21. 38 x 0.9 = ______ (34.2)
22. 66 x 0.5 = ______ (33.0)
23. 68 x 0.5 = ______ (34.0)
24. 11 x 0.5 = ______ (5.5)
25. 2.5 x 2 = ______ (5.0)
26. 6.4 x 2 = ______ (12.8)
27. 8.5 x 8 = ______ (68.0)
28. 8.9 x 4 = ______ (35.6)
29. 4.2 x 9 = ______ (37.8)
30. 9.6 x 2 = ______ (19.2)
31. 0.2 x 6 = ______ (1.2)
32. 7.4 x 2 = ______ (14.8)
33. 8.7 x 5 = ______ (43.5)
34. 6.0 x 7 = ______ (42.0)
35. 3.4 x 9 = ______ (30.6)
36. 4.8 x 5 = ______ (24.0)
37. 7.4 x 7 = ______ (51.8)
38. 5.3 x 5 = ______ (26.5)
39. 3.2 x 6 = ______ (19.2)
40. 0.3 x 9 = ______ (2.7)
41. 2.6 x 1 = ______ (2.6)
42. 0.9 x 8 = ______ (7.2)
43. 2.2 x 3 = ______ (6.6)
44. 1.7 x 7 = ______ (11.9)
45. 9.6 x 5 = ______ (48.0)
46. 9.7 x 8 = ______ (77.6)
47. 0.9 x 9 = ______ (8.1)
48. 8.0 x 8 = ______ (64.0)
49. 0.8 x 3 = ______ (2.4)
50. 1.2 x 3= ______ (3.6)

6.4.3 Decimal Division Worksheet 3

The following problem sets are multi-digit dividend values into single-digit dividers. Answers are provided in parentheses (…).

1. 0.7 ÷ 3 = ______ (0.23)
2. 0.9 ÷ 5 = ______ (0.18)
3. 0.3 ÷ 2= ______ (0.15)
4. 0.2 ÷ 5 = ______ (0.04)
5. 0.5 ÷ 4 = ______ (0.13)
6. 0.3 ÷ 6 = ______ (0.05)
7. 0.7 ÷ 2 = ______ (0.35)
8. 0.4 ÷ 5 = ______ (0.08)
9. 0.6 ÷ 7 = ______ (0.09)
10. 0.9 ÷ 6 = ______ (0.15)
11. 0.4 ÷ 6 = ______ (0.07)
12. 0.9 ÷ 3 = ______ (0.30)
13. 0.9 ÷ 8 = ______ (0.11)
14. 0.5 ÷ 2 = ______ (0.25)
15. 0.5 ÷ 7 = ______ (0.07)
16. 0.5 ÷ 9 = ______ (0.06)
17. 0.6 ÷ 4 = ______ (0.15)
18. 0.2 ÷ 8 = ______ (0.03)
19. 0.7 ÷ 6 = ______ (0.12)
20. 0.9 ÷ 3 = ______ (0.30)
21. 0.7 ÷ 7 = ______ (0.10)
22. 0.1 ÷ 2 = ______ (0.05)
23. 0.7 ÷ 8 = ______ (0.09)
24. 0.2 ÷ 2 = ______ (0.10)
25. 0.1 ÷ 9 = ______ (0.01)
26. 2.5 ÷ 3 = ______ (0.83)
27. 6.4 ÷ 5 = ______ (1.28)
28. 8.5 ÷ 7 = ______ (1.21)
29. 8.9 ÷ 2= ______ (4.45)
30. 4.2 ÷ 4 = ______ (1.05)
31. 9.6 ÷ 6 = ______ (1.60)
32. 0.2 ÷ 2 = ______ (0.10)
33. 7.4 ÷ 5 = ______ (1.48)
34. 8.7 ÷ 7 = ______ (1.24)
35. 6.0 ÷ 8 = ______ (0.75)
36. 3.4 ÷ 6 = ______ (0.57)
37. 4.8 ÷ 2 = ______ (2.40)
38. 7.4 ÷ 3 = ______ (2.47)
39. 5.3 ÷ 8 = ______ (0.66)
40. 3.2 ÷ 2 = ______ (1.60)
41. 1.3 ÷ 6 = ______ (0.22
42. 2.6 ÷ 9 = ______ (0.29)
43. 0.9 ÷ 4 = ______ (0.23)
44. 2.2 ÷ 2 = ______ (1.10)
45. 1.7 ÷ 5 = ______ (0.34)
46. 9.6 ÷ 6 = ______ (1.60)
47. 9.7 ÷ 3 = ______ (3.23)
48. 0.9 ÷ 7 = ______ (0.13)
49. 8.1 ÷ 2 = ______ (4.05)
50. 1.6 ÷ 9 = ______ (0.18)

6.4.4 Decimal Division Worksheet 4

The following problem sets are multi-digit divisor values into single-digit dividers. Answers are provided in parenthesis (…).

1. 1,083 ÷ 0.7 = ______ (1,547.1)
2. 5,067 ÷ 0.6 = ______ (84,45.0)
3. 1,664 ÷ 0.2= ______ (8,320.0)
4. 5,931 ÷ 0.9 = ______ (6,590.0)
5. 7,664 ÷ 0.2 = ______ (32,320.0)
6. 8,045 ÷ 0.7 = ______ (11,492.9)
7. 9,779 ÷ 0.8 = ______ (12,223.8)
8. 5,212 ÷ 0.2 = ______ (26,060.0)
9. 448 ÷ 0.9 = ______ (497.8)
10. 468 ÷ 0.8 = ______ (585.0)
11. 4,854 ÷ 0.4 = ______ (12,135.0)
12. 1,816 ÷ 0.3 = ______ (6,053.3)
13. 5,184 ÷ 0.9 = ______ (5,760.0)
14. 4,953 ÷ 0.7 = ______ (7,075.7)
15. 96 ÷ 0.4 = ______ (240)
16. 1,190 ÷ 0.1 = ______ (11,900.0)
17. 3,984 ÷ 0.9 = ______ (4,426.7)
18. 2,030 ÷ 0.3 = ______ (6,766.7)
19. 6,150 ÷ 0.7 = ______ (8,785.7)
20. 5,450 ÷ 0.3 = ______ (18,166.7)
21. 7,392 ÷ 0.2 = ______ (36,960.0)
22. 2,988 ÷ 0.2 = ______ (149,40.0)
23. 6,624 ÷ 0.7 = ______ (9462.9)
24. 2,940 ÷ 0.2 = ______ (14,700.0)
25. 3,042 ÷ 0.6 = ______ (5,070.02)
26. 1,693 ÷ 0.1 = ______ (16,930.0)
27. 8,055 ÷ 0.9 = ______ (8,950.0)
28. 8,480 ÷ 0.9 = ______ (9,422.2)
29. 4,032 ÷ 0.5 = ______ (8,064.0)
30. 4,512 ÷ 0.4 = ______ (11,280.0)
31. 5,015 ÷ 0.4 = ______ (12,537.5)
32. 7,152 ÷ 0.9 = ______ (7,946.7)
33. 2,936 ÷ 0.2 = ______ (14,680.0)
34. 5,664 ÷ 0.9 = ______ (6,293.3)
35. 5,010 ÷ 0.6 = ______ (8,350.0)
36. 1,060 ÷ 0.2 = ______ (5,300.0)
37. 8,104 ÷ 0.2 = ______ (40,520.0)
38. 1,404 ÷ 0.4 = ______ (3,510.0)
39. 82,893 ÷ 0.1 = ______ (828,930.0)
40. 916 ÷ 0.7 = ______ (1,308.6)
41. 2,905 ÷ 0.3 = ______ (9,683.3
42. 6,416 ÷ 0.3 = ______ (21,386.7)
43. 1,330 ÷ 0.8 = ______ (1,662.5)
44. 5,568 ÷ 0.1 = ______ (55,680.0)
45. 5,395 ÷ 0.6 = ______ (8,991.7)
46. 6,020 ÷ 0.9 = ______ (6,688.9)
47. 6,579 ÷ 0.5 = ______ (13,158.0)
48. 4,224 ÷ 0.5 = ______ (8,448.0)
49. 2,444 ÷ 0.5 = ______ (4,888.0)
50. 891 ÷ 0.9 = ______ (990.0)

Appendix

INDEX of KEY WORDS

Index of Key Words.

The following terms alphabetically listed and are found in this book. The values following the hyphenation represent the chapter number (integer) and the page (decimal). Page in bold address the best understanding at the indexed term.

A

B

C

D

F

G

H

O

P

R

T

Made in the USA
Monee, IL
05 November 2026

67807073R00146